E≠mc²

(E is not equal to mc²)

W. J. McKee

PUBLISHER:

William James McKee
311 Paradise Drive
Tiburon, California 94920
http://www.relativity-lightspeed.com

Cover Design by W. J. McKee
Cover Image by Robert Gendel
Cover Layout and Interior Illustrations by Paul Street

ISBN: 0983345058
ISBN 13: 9780983345053
LCCN: 2011902543

ILLUSTRATIONS

Illustrations by Paul Street, Lake Worth, Florida

COVER IMAGERY IS OF THE M31 (ANDROMEDA) GALAXY.
PHOTOGRAPHED BY ROBERT GENDLER

DEDICATION

The dedication of this work would be ordinary were I to place it at the feet of the great scientists gone before us, and I therefore do so dedicate it, but it is necessary to also recognize those in my immediate life experience and declare my undying appreciation to my father William James McKee, Senior, my wife Barbara Lee, and my youngest sister Eleanor Marie for the early experiences and education, the strength of curiosity and determination, and the drive to accomplish what others see as daunting. Thanks and high recognition go to our illustrator, Paul Street. It must also be dedicated to you readers who have the spark of interest and desire to understand the world and universe around you. I must take my hat off to such individuals; always press onward.

William James McKee

FORWARD

This work is written with the bright inquisitive non-scientist reader in mind. While there are sections that are complete, with sufficient mathematics to formalize the arguments within, there is also extensive text to lead the uninitiated reader through what some would fear as difficult mind-torture. This book looks at fundamental theories, including some background results, and in the final chapters moves up to present theories and present alternative possibilities.

Decades have passed and strides have been made in the sciences: modern equipment and observations, voluminous data gathered, and further calculations and speculations made. We have relied upon acceptance of obscure theories and those of prominence without adequate skepticism and study. It's time to begin anew and investigate other possibilities or confirm those set down in the past.

The reader will find much food for thought and, although we can't perform some of the experiments put forth at this point and time in our development, we can at least begin together to move in that direction. All persons with the least bit of curiosity will find interesting points brought forth with sufficient clarity to present them with new perspectives about the universe we reside within.

One may look at this work as a primer on some fundamentals in the field of astrophysics, and that includes many of the historical points accomplished by many of the great scientists that have gone before. We will revive the arguments of the emission theory of photon existence (also known as the ballistic approach) with a new approach and persuasive logic. With the march of time, we have the pleasant advantage of being able to join much of the accumulated works done in this area and utilize more modern technological methods and equipment in search of data in support thereof.

We, as humans situated on one small planet in one small galaxy in this expansive universe, should stop and consider the amount of information and data that we don't have in order to try and understand our surroundings. We can only move along step-by-step and try to apply our findings with the events and objects we observe, keeping in mind that the explanations we come up with may very well change over time as more data is gathered and knowledge acquired.

Our meager attempts at explanation of any events are sophomoric at best and border on folly, but we must by our nature throw ourselves head first into such endeavors. We have, therefore, moved down the only path possible, and that is to take our information, data, observations, newest equipment and apparatuses, along with logical reasoning, and move slowly toward the **true** answer in the quest for the answers to the mysteries of the universe.

Much has been said about the various photon theories, and some have been dismissed, such as the emissive theory. This is partially due to an interpretation of experimental results that may or may not be correct. The idea of a preferred frame of reference and the presence of a cosmological aether seems a tenuous basis for formulating a strict opinion and then dismissing those arguments that disagree out-of-hand.

The author will often repeat or restate ideas and principles in order to tie various thoughts together. This work moves to reinstate the emissive theory as a valid, if not controversial, theory. This is not intended to bore the advanced reader but to demonstrate relationships in the variety of ideas that are so intertwined in this field of study.

All of us who have driving interests in the way things work and the underlying mechanisms supporting what we observe can always make contributions to the science even if it is limited to generating a spark in others.

It is strongly suggested that the reader take some time and look at the English translation of Professor Einstein's 1905 presentation on Special Relativity Theory (SRT), as translated and presented by R.W. Lawson in 1920. (http://www.bartleby.com/173/). Much of the work cataloged therein has been generated in response to this publication by Professor Einstein.

The author has, on occasion, used terms such as "he" in referring to a nebulous person or individual. This is done without intent to assign gender but only for smoothness in relating ideas. No discrimination against the female personage is in any way intended.

INTRODUCTION

For millennia, mankind has looked to the sky in wonder. We have collectively searched, studied, and imagined just what was out there. Many of us in the past have imbued the universe with mysterious origins; a lot of us embrace a "big bang" theory. We have had numerous learned men considering light and speed relationships, physics, celestial bodies, gods, and the other life forms that most certainly exist out there.

We have gone through periods in our learning where the earth has gone from being the center of the universe to a mere speck in a small galaxy. There have been those who had the sun and other planets revolving about the earth. Some thought that the earth was flat—and it appears that some still do! Some have thought the sun and moon to be gods.

As long ago as 1600 B.C., early man was looking upward and casually taking stock of what was going on up there and eventually finding out how it affected their lives down here. They became aware of the seasons and the phases of the sun and moon in their respective treks across the sky. Tablets were made, with early writing about the celestial goings-on.

Objects were made, showing that early man knew about such things and, even though they didn't have writing skills yet, they put their ideas down skillfully in drawings, symbols, and beautifully crafted objects that were obviously very important to them as a culture.

IT'S A ROUND WORLD AFTER ALL

We've been on a constant journey thus far in learning about ourselves as well as our celestial neighbors. When you finish this book, you will have been introduced to new ideas about our universe and its workings. Don't be too quick to accept or reject what you read; it is intended to spark your thought process and thereby get you to join all of us with your opinion or at least to join conversations with your interests in and perspective of our universe.

As we have found over the centuries, many of our concepts and presumptions have been shown to be somewhat flawed. This, of course, is the result of trying to define our observations within the limits of our instruments and imagination at the time.

We continue to improve our instrumentation and build upon the knowledge base of our predecessors to move steadily forward. Eventually we will have our shortcomings or flaws in our conclusions also exposed for an even more accurate picture of our cosmos. Perhaps this is the real never-ending story.

It has been considered and now shown that much of the great work of the past, while of exceptional value to our progress, was nonetheless erroneous in part. Some of these errors have been sophomoric in nature and some very obscure and not obvious until decades later.

Many of the past masters were revered so highly that many followers would simply accept their conclusions without adequate understanding and skepticism or appropriately questioning. I'm sure the same fate will find the work of all of us eventually, as our knowledge base grows and our technology improves to support ever more sophistication in experimental efforts and the quest for knowledge.

One of our problems as human beings is a willingness to accept suppositions and conjecture as correct or valid without adequate reasoning or proof. If a charismatic "maestro of thought" proposes a new concept or explanation to a problem, we listen and, through the professing of lengthy arguments, get caught up in the stream of logic (or illogic) and become confused or entrapped; we think that what we are hearing is obviously correct because it appears plausible or intuitively obvious, so we accept it. Don't do that!

Read, think, question, argue, study, understand, or at least struggle to do so, and then converse with others who both agree or disagree and test your understanding. Question the work presented herein. Read with skepticism if you must but read it and at least consider the possibilities.

While I expect the following pages to make you think more about our situation, and while I seem to discredit many ideas of some past great minds, I don't mean to dismiss their efforts.

I have great respect for them and only wish to put forth a different approach, a new possibility, perhaps just an extension of some earlier work just as those who will come after us will do for us.

This work is not intended to be a serious scientific work on the subject matter but instead one that will engage the general student population that may be somewhat unfamiliar with the notion of physics and its many branches of study. While some specific references to past works and some mathematics are included, they are here for the intermediate student and to clarify a particular concept, not as rigorous proofs.

With the Information Age well under full sail, we are able to journey the world over in search of information on any subject matter we want, in volumes so vast that the task becomes foreboding without careful discipline to limit one's search parameters. One can easily find oneself spinning off on a tangent, attracted by much of the wonderful information sources available out there. The path of study is often not a straight one.

We can look at past experiments, read about the pros and cons of each argument, and, with such a treasure trove of information like never before, begin to piece together ideas and hypotheses on whatever subject we wish to address. Once an introduction to a particular idea or field of endeavor is formulated, one finds it necessary to stray a bit in order to glean some of the information from underlying work in complementary areas of study done by others.

Areas of scientific study are so interlinked that no one person can hope to study or learn the full depth and breadth of all of them, let alone become a master of even one narrow segment. In studying the works of Newton, Maxwell, Planck, Michelson and Morley, Doppler, Fizaeu, Lorentz, and others, it became clear that many questions existed with not quite credible enough answers to bring them all together in a cohesive and elegant solution.

Einstein's work on special relativity was troubling. When we had to look at space-time with mass perturbations, linear compression, time dilation, and a preferred frame of reference from which the speed of light was measured, something appeared seriously wrong.

In such a realm, the ultimate speed limit was thereby deemed to be defined for all other frames of reference. This wasn't sensible, intuitive, or reasonable, and one's mind couldn't quite wrap comfortably around these concepts.

What wondrous ideas mankind had come up with over the many decades of study! What ingenious ideas, devices, experiments, thoughts, and carefully considered analysis along the way. It became obvious that there have been truly intellectual giants among us. The impulse of bowing down seemed somewhat appropriate. As we all gain in knowledge, we become more and more humble.

Without doubt, amongst the billions of other galaxies and the billions of solar systems within them, we will eventually find numerous populations of living beings. No doubt, as well, some will be far more advanced, and some much less so, but certainly in existence. We have only the opportunity to use mankind as a benchmark by which to judge the others, not expecting ourselves to be either more or less advanced than those we happen to encounter.

It's a shame that we haven't learned to travel the stars yet in a way that makes intergalactic journeys available to us. We could certainly benefit from contact with others. As we grow as a species, one would hope that we can also learn to control ourselves to the point wherein we can work in unison towards commonly advantageous goals and begin venturing out—going slowly at first, certainly, carefully, with curiosity, enthusiasm, and a tenacious determination, seeing, watching, testing, and studying what wonders we'll encounter.

Like many other thoughtful people who have gone before us, perhaps we can take a journey through the universe by way of thought and see what we find. Some surprises are there. Many interesting things to see are there. Perhaps some of the answers we've been looking for are there as well.

ALIEN IDEAS

We're going to have a good time of it. I wonder if we'll find some exciting repositories of knowledge, wonderful emerging populations like ourselves, or answers put aside for those interested enough to venture out. We'll zip along with photon streams, meet other beings, conduct a few simple experiments, and generally have a great time.

I'll tell you what. Let me invite you along on a flight of fancy on a beautiful afternoon; we'll venture out to the stars. We'll take my new spaceship—built it myself. It will hold us comfortably, and we'll go out and see the wonders of the universe together.

BUILT IT MYSELF

In this journey, we can see some of the awesome revelations that come into play as we accelerate up to, and then beyond, the speed of light. So up we climb onto the shoulders of some of the giant intellects that have gone before us, without whom we would still be slowly gliding along in a terrestrial vehicle.

Please understand and, by way of explanation, accept the acknowledgment that much of this work is not original to me but accomplished by many brilliant colleagues over the decades. My idea is to try to tie some of the strings together into a comprehensive bundle, to stir the thought process and spark discussion on the topics.

Most respectfully submitted for your enjoyment,
W. J. McKee

TABLE OF CONTENTS

OUR JOURNEY BEGINS

Our most primitive ancestors in their wanderings had little recognition of the vast cosmos above. Undoubtedly, they were, however, highly conscious of the presence of the stars, the sun, and moon because of the light and warmth provided. Over the centuries, mankind has looked to the heavens with increasing wonder.

The ancients assigned powers and mystical properties to the stars and constellations. Astrologers built a cult-type reverence and formulated a highly complicated for-mulae, whereby they believed, and many still believe, that they could tell the complete background, the future capabilities, and events for any individual just by know-ing the moment of birth and reading the positions of stars and constellations.

Primitive people learned to pay great attention to the chang-ing seasons and, as their powers of observation and logic ma-tured, found that the turning of the earth, the different phases of the moon, and the annual climate changes were all becom-ing familiar and were somewhat predictable. They could use them to judge the best time to plant and harvest crops.

Those near the seas could count on the tides for assistance in fishing and launching of boats, as well as less pleasant methods of torture. Eventually they associated the tides with the movement of our mysteriously regarded nearby compan-ion satellite, our moon.

Many ancient societies built observatories and special shrines involving the annual changing seasons. The Aztec civilization developed a calendar and constructed observatories.

Some ancient New World native American cultures built their dwellings with an eye to the setting and rising sun. Consider monuments like the English Stonehenge, the Egyptian and Aztec pyramids, and the Inca observatories; in the primitive western hemisphere, extensive ground-based images were designed and laid out upon the desert and hills that would only be viewable from a flying object or the gods looking down from the heavens.

Humankind very early on found that fire provided warmth and light as well as great pain when approached too closely. Eventually they found ways to handle it and understand its ways. No thought was given at that time as to the science behind the flames. Galileo looked through his handmade telescope at the moon, stars, constellations, and planets. His efforts in trying to define how the cosmos worked, even in direct competition with powerful religious theologists, were insightful and daring. He took notes describing his observations, and one conclusion was that the Earth was not the center of all things as the theologists would have had the people believe.

LOOK TO THE STARS

For centuries, men have made up names for the numerous constellations and stars. Nebulae were observed with the curiosity of a child, and mankind named them as well. We, as humans, have envisioned the heavens to be the seats of godly power. Tales were constructed about the many gods that roamed the skies and earth with their mystical powers and heavenly abodes.

We are still making amazing discoveries through use of extraterrestrial observation instruments such as the Hubble telescope. We have sent out various satellites to have a look around, especially at our moon and neighboring planets. We have devised, designed, and fabricated sophisticated listening devices in search of alien civilizations that may be out there.

3

Scientists have been interested in and studied space, the beginnings of the cosmos, and how gravity works with the large massive bodies. They have watched how the planets moved about and then used their observations to formulate various rules that brought their observations into our more familiar world and understanding.

The ideas of an aether, an invisible flowing medium, as a ubiquitous pervasive part of all the universe was put forth but found to be improvable, although numerous experimenters made bold attempts. A fascination with black holes and dark matter is now in vogue and the subject of much consideration and conjecture.

Many ideas concerning light have provided impetus to numerous investigators throughout the centuries.

We have, over the years, found ways to harness electrical power to generate light using objects such as:

- wire filaments
- carbon rods that move closely together and pass an electrical arc between them
- materials that fluoresce when bombarded with an electron beam
- small "solid state" devices, such as light-emitting diodes.

Man eventually realized that light was a medium that could transfer image information to the observer and thereby was a tool to aid in further investigations of the world around him. It eventually became apparent that light also had a component of energy; in moving about in the sunlight, one felt the warmth received by the illumination. What was this light? From where did it come, and why did it provide warmth?

It was obvious that the major provider of light for us on this planet was our own sun. It was not considered at first that all the stars in the sky that were visible to the casual observer were also "suns." So here we are, circling around a large burning gaseous furnace that is spilling out vast amounts of radiations, including particulate matter and an unimaginable volume of photons.

Was "light" made up of particles or waves? Was it an electromagnetic form of radiation? Why did it warm objects it also illuminated? How did refraction work? At what speed was light really traveling through space? How about as it passed through other mediums such as glass, water, oils, vacuums, and such? So many brilliant scientists have worked on these ideas for the past centuries that the list reads like a Who's Who of the Scientific Community.

Today we still scan the skies for remnants of possible communication or signs of intelligent life outside our own planet. We watch the heavens and see unidentified flying objects. We view videos that show unusual and fast-moving objects, some of which are yet to be fully explained.

We are certainly a curious lot, and eventually humankind may come to better understanding and wisdom that will make us contemporary folk seem quite primitive to those who follow centuries, or perhaps only decades, from now. Technology and information, new findings, and capabilities are rapidly pushing forward our race's ability to proceed ever faster and faster.

In the text of this work, we cannot pursue all of the characteristics of light and interactions with photons. There is too much underlying science and history to bring into a single volume of work. We, instead, can acknowledge the substantial body of work done in these fields as significant, if not wholly correct, and applaud those who have gone before us with curious minds.

In this volume, we must refrain from delving too deeply into the sciences of thermodynamics, gravity, electromagnetic theory, optical refraction, human pathology of vision and visual studies, and so many others that are intimately related to the subject matter herein. Such connectivity with the other sciences would fill a library by itself and is far beyond the scope of this work.

Some time ago, several renowned scientists were speculating about the cosmos, velocity of light, the aether in the space vacuo, dimensional contraction, as well as time dilation, among other esoteric subjects. There were experiments with heated black bodies, spectrometers, electromagnetic waves, and categorizing various types of radiations.

There have been years of questions and speculations as to whether light worked as waves or particles. Behind all of this are the underlying fields of mathematics, logical deduction, and the ability to communicate our thoughts and findings to others.

Many scientists have steadfastly held on to the theories put out by some well-known experimentalists and the interpretation of their data. Thought experiments have been embraced as factual, and their results, while unconfirmed, have become the standards upon which astrophysicists have based many of their theories and ideas. Some have been termed laws and are treated as such.

Someone has been having fun with the pliable minds of others, indeed many others, and reasonably intelligent ones in some cases. Perhaps a fanciful and enlightening trip will assist us in further understanding the cosmos and its workings.

Let's take a brief trip out into space and have a look around as others have done in the past. We'll take our spacecraft deep into the cosmos. Nothing else is close enough to us to affect our craft with significant gravitational fields. We're "light years" away from the nearest celestial body.

We can use the idea of thought experiments and our imagination to help us. We get aboard and are off, as our journey begins.

Our little spacecraft needs no fuel, can accelerate gently or instantaneously to whatever speed we would like, and travels without physical restrictions or limitations. We won't be bothered by collisions with space debris or get overly warm.

We have windows everywhere and can easily see to the front, rear, and all around as we go speeding through the cosmos. We have green running lights fitted fore and aft for safety. In our special instrumentation panel, we have interesting equipment and other items to permit some basic measurements to augment our limited human senses. We get on board and we're off. As we gently increase our speed, we enjoy the sights.

Perhaps we should visit the Orion Nebulae or the Eagle. No matter, let's just have fun!

As we accelerate gently up to and then pass the velocity marked c on our speedometer, we have not been observing the view out the rear of the craft. Now, for the first challenge, we'll momentarily reject the well-practiced version of the finite and limiting value of "c" (the speed of light) as put forth by A. Einstein a little over a hundred years ago. We'll use a nice round number approximation of 300,000 M/s (Meters per second) for c as a nominal, mean value for convenience.

We see to the rear that many of the luminous objects are diminishing in brightness and eventually can no longer be seen. What's happening? We'll concentrate on looking at our surroundings and trying to understand how the "laws" of physics really work out here.

The characterization of light is of paramount importance in our observations, and understanding how it behaves will be the main object of our focus for a while.

WAS THAT A PHOTON WE JUST PASSED?

As the sources of light to our rear recede, they seem to begin to disappear. In their transition, they initially demonstrate what appears as a spectral shift from the apparent white, then through red, then infrared, and then gone from view. (For this portion of our talk, we'll use the human "photopic" visual range of wavelengths as our detector until other means are introduced).

But wait—if Einstein was wrong about the relative speed of light, we find that there are a few stars and clusters still visible to us even though the large majority of them have disappeared! Yes, there would be some! They are the stars, galaxies, clusters, and other bodies that are moving substantially in our direction while emitting their light.

They would have been blue-shifted when we were stationary in the local space partition (LSP), but now they appear as red-shifted because of our ever-increasing velocity moving away from them. Soon, as we continue to increase our velocity, we see all of them eventually disappear from our view.

If we deviate from the view that Einstein proposed (that defined a limiting speed of light to be a constant and not dependent on, nor tied to, the velocity of the emitting source), and we formulate the idea that the speed of light was strictly relative to and linked firmly to its source of emission, we would then algebraically add the velocity of the source relative to the sensor and the emission velocity of the photons being emitted from that source.

This is, of course, what actually occurs and what we've been observing; it is also a Newtonian, ballistic, or emissive view of such ideas, as well, but we'll get into that a bit later.

Let's lay a bit of groundwork here to embrace some of you who may have an apprehension about our trip and observations. Suppose that when we are riding along, we come across some light beacons that are strung out guiding us along a path in space. Let's suppose that these beacons are like space buoys and are stationary in the local frame of reference of the LSP.

Let's define the LSP for our convenience as that frame of reference that is conveniently and essentially stationary within the local space, wherein there is no apparent translatory movement with respect to distant observable celestial bodies visible to the casual observer over a short period of viewing time.

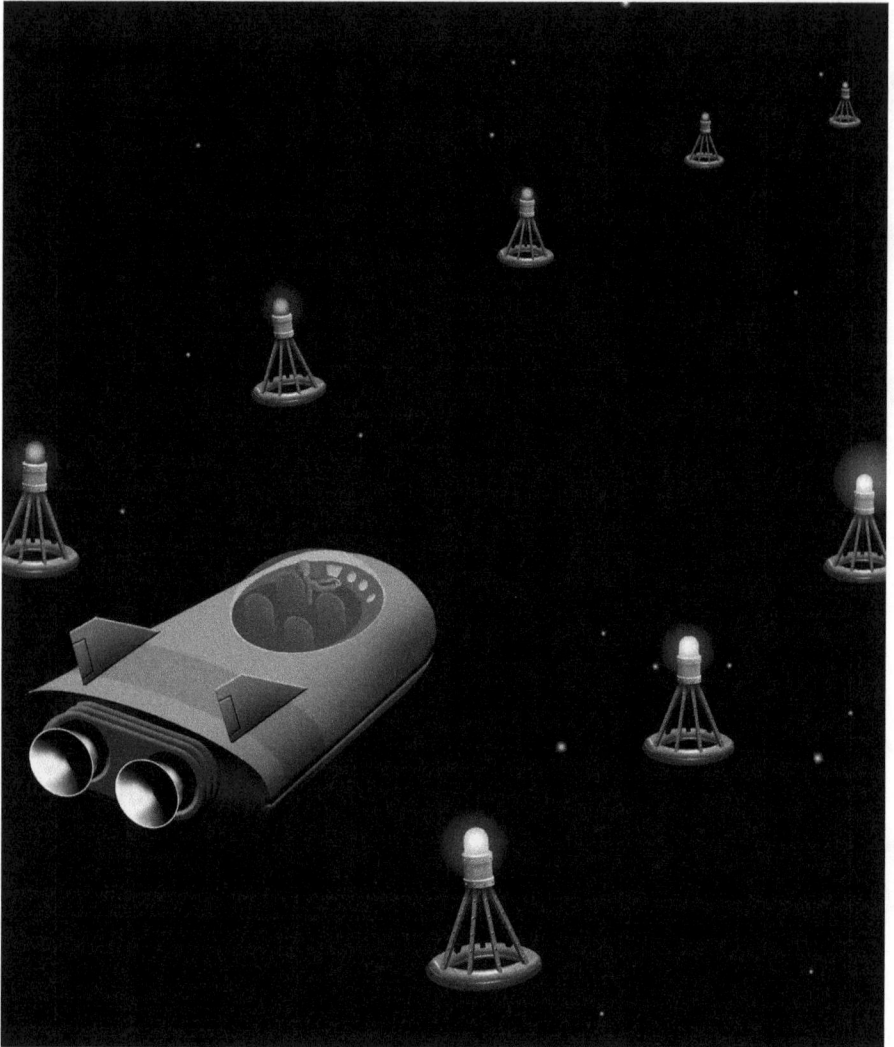

SPACE BUOYS

Rotational movement about an imaginary axis where the observer resides is not considered as a part of this convenient definition, for simplicity of explanation. This loosely describes a situation similar to that we experience as living beings on our planet Earth as we look out at the vastness of the universe. We could consider our solar system as our LSP.

The expansion of the universe as we now know it, with vast rotational excursions of galaxies and other celestial bodies, escapes casual observation, and we exist in our own LSP. Since the term includes the word *local*, it is limited for our purposes to space that is within, say, a few light-years or so. While this constraint seems arbitrary, it is set as a convenience to constrain further tangential wanderings of our considerations.

The idea of this partition is required to allow the relative measurement of the action of other objects within the observable domain of the traveler or to the residents of that partition that observe any goings-on that may transpire in that particular region of space. Back to the buoys. Suppose that the buoy lights output a steady amount of photons, and let's also suppose that the light beacons on our left are strictly monochromatic in their emission how about bright green (as seen by the observers resident in the local space partition) for reasons we'll discuss later.

The beacons on the right are of a very broad spectral character; all wavelengths are being emitted at exactly the same intensity. (For the purposes of this conversation only, let's assume that the human observers' eyes will conveniently take these equal levels of all photon spectra and combine them to perceive white light).

Now let's look at why we choose the single color on the left side buoys. We chose green because it is in the vicinity of the central wavelength of the human photopic range. As we accelerate up to the speed of light, we will see a continuous spectral shift proportional to our speed, relative to that source. If the light source is strictly monochromatic, we encounter such a pronounced color shift until we can no longer see the light source with our eyes. The spectral shift has moved too far for our eyes to see it, but with our fantastic, special, broad spectrum, spectra-photometer we can observe. as the shift continues with our ever-increasing velocity, as it passes beyond the human photopic range.

If we are moving toward the photon source, this shift would look to our eyes as a gradual change from the initial green to blue, then violet and then into ultraviolet and beyond the range of our vision. If we observe these same buoys out our rear window and watch them recede into the black of open space, as we accelerate up to and then beyond the speed of light, we'll see them appear to emit light with a shift toward the red until it passes into the infrared and then again beyond our ability to see them.

Imagine now that the light source (like the beacons on the right side) was made up of all frequencies of the light spectrum, well beyond the photopic region commonly detectable by our eyes. Our eyes will see an increasingly bluish spectral appearance! Why? Well, if the photon source comprises all possible photon energy levels, all equal in amplitude and number, the lower energy (infrared and beyond) photons gaining speed will move to replace those above in a continuum, thereby always giving a gently increasing energetic beam steadily moving toward the ultra violet end of the spectrum in appearance to the observer's eyes.

To our rear, we see the energy-detecting meter (looking solely at the human photopic photon range) taking an ever-decreasing reading since the higher energy photons pass the upper limit of our vision (ultraviolet) and are continually replaced by lower energy photons at the infrared spectrum point, coming in to replace them on their march through the visible spectrum (and thereby through the limited energy range).

Now, we know that most celestial sources emit (or reflect) light (photons) with a broad spectrum. The resultant stream of photons, emitted with an apparent whitish mixture and having a number of characteristic peaks within the spectral band, have been found to shift as expected through blue, then violet, then into the ultraviolet and beyond, as the source and we observers move ever faster toward each other.

The spectral "fingerprint" of that particular light source will afford us the ability to accurately determine this spectral shift relative to the velocity differential between the emission source and the observers' sensor.

We can recall observations of such as emissions from particular galaxies moving away from us (and the observing sensor) and clearly point out the attendant spectral shift. This shift is commonly defined by the use of a Doppler formulation, albeit not precisely so. This Doppler shift is well known and instrumental in demonstration of the variation in sound wave frequency as the source of the sound is moving either toward or away from the observers' position.

One significant difference in the application of the Doppler analogy between sound waves and light is that the sound waves require a medium through which the sound waves are transported. The photons making up what we call light are actually particulate and, as such, do not require such a medium.

Many people view these photons as a part of the electromagnetic spectrum of radiations. Maxwell's equations have been linked to them, and much of our historical formulations and theories are linked to this approach.

There has long been a controversy between the camps of believers in whether the light travels in waves or is particulate in nature. Given the above evidence, including contributions by such men as Newton and Planck, we must conclude that light is photons on the move and that they are in such numbers that they, like water molecules, combine and can form waves on a macro basis. With this particular analogy, one can understand how many of the questions continually brought up in discussions about the characteristics and behavior of light can be better understood.

It becomes essential at this point to be cautious in categorizing light as electromagnetic. While electromagnetic waves behave in a manner similar to some light waves, they are not in the same family. Photons passing along an optical fiber do not cause a magnetic disturbance in the vicinity of the optical fiber.

One must look at the idea that electrons moving in a copper wire can cause such perturbation, even at low velocities of movement, while photons moving along at much greater velocities have no such effect associated with them.

A different sort of radiation indeed!

2

DISTANCE AND TIME

One of our problems in both conducting our experiments and discussing the results with others is the ability to measure. We first must define what measuring is, come up with units of measure, formats, and procedures for the various types of measurement, and finally devise the system of coordinates that will allow us to display the results in an orderly and consistent format.

Let's consider a coordinate system so we can begin to discuss our place in the universe and to reference others. Many others have constructed these types of coordinate systems by having three planes at right angles to each other; the planes all meet at the point of origin. The Cartesian coordinate system is the de facto standard and easily understood by us. This is also a completely valid approach, but further understandings and constraints are in order.

What do we mean by three planes at right angles? (See: Euclidean Space) What do we mean by three? We must begin with a means to devise a numbering system in order to begin quantifying our information. We need a means to measure lengths, passage of time, angular measurement, and measurement of all the various physical parameters that may be associated with the observations we make.

It is required to begin with very basic definitions about those quantities and properties that might become useful in our attempts in this area. Many of the tools we'll use are already devised, and we simply adapt them for our particular situation. The Cartesian coordinate system is primary in importance.

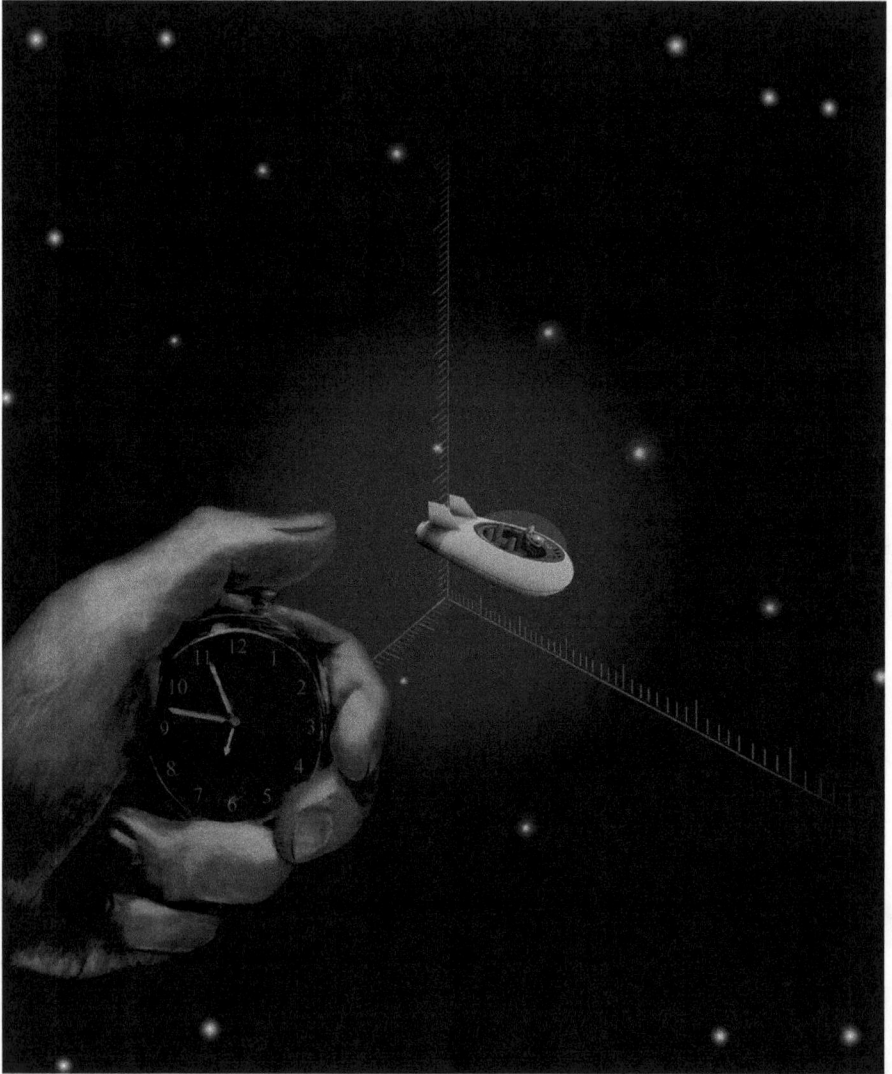

TIME AND CO-ORDINATE SYSTEM

We could fill volumes of texts with the variety of possible approaches to define and construct the means for fixing a numerical system, a dimensional methodology, a subsequent coordinate system, and other means of measurements far beyond the point of this book. Without doubt, the reader can certainly appreciate the vast amount of effort applied to these tasks throughout the centuries.

Trying to synopsize these efforts and results would not do them justice and will not be attempted herein. Now we can begin to say, albeit in the most crude and fundamental terms, where other celestial objects are located with respect to our coordinate system for the time being. Imagine now that other space travelers are some moderate distance away and set up identical coordinate systems so they can also define their place in the cosmos and that of neighboring celestial bodies.

Nothing in these newly defined imaginary coordinate systems includes any rotational component for any axis. Without doubt such rotation exists and will, over a length of time, perturb the results of our experimentation. If we take precautions to make our experiments and observations within a brief time period, the results can be considered to be valid and will relegate the existing rotational components to being inconsequential for the moment.

The simple results of the thought experiments we attempt to undertake will change over time, but the precepts and logic underpinning them will remain valid. A correlation function over the future decades may be worked out in order to maintain the validity of our actual results.

Let's make up a marvelous communication device (system) that can communicate with others without any time lag between the time of transmission and reception. No time factor for the travel of these transmissions, just instantaneous communications.

Now let's have the communication device also contain a universal translator so everyone using it can understand everyone else. Now we send out a transmission and ask if anyone is out there to talk to. We receive instant responses in the affirmative, and we can begin our chat.

When we begin talking, we suddenly understand that while we can communicate with and understand what the other fellows are saying, we can't tell them about our perception of the universe because we don't speak the same systems of length, time, power, energy, or color among the myriad other parameters.

UNIVERSAL COMMUNICATOR and TRANSLATOR

One thing that will help is that all of our spacecrafts are outfitted with brilliant green lights so others can at least see them in the darkness of space. Again, green has been chosen because it is near the center of our human photopic spectrum, which advantage will become obvious as our travels evolve. Let's also decide that all of our spacecraft are of the same design and construction and are, in fact, identical in every way.

For a quick way to begin discussions with our alien friends in trying to describe ourselves, our planet, our understanding of all things in the realm of physics in the universe, we need to set a few points of basic metrics. How can we possibly begin to make sense of a common terminology? Fortunately, we can use the standard number system since it is universally sound. There are no arbitrary basic assumptions or units of measurement, just numbers from minus to plus infinity with zero in the center of origin—our basic number line learned in primary school.

Now we find that our cosmic friends are scientifically and technically savvy. They have access to lots of scientific equipment and facilities in their locale, so we can begin to discuss and understand their systems versus ours. Perhaps we should begin with our concept of measuring units of length or time or perhaps color! We have an interesting problem here! They don't have any meters or inches or seconds or color wavelengths or anything else that is so familiar to us and our measuring systems. We must make a determination as to which parameter we wish to first define in universal language.

Okay, now let's talk about a circle or sphere? We tell them that in our studies of geometry, we use 360 degrees to section off a circle with subdivisions of minutes and further minor divisions of seconds. The trigonometric functions are clearly known and understood, and now we can easily converse in geometry.

How do we discuss length? How long is an object? What item or object can we refer to that is going to be the same form for them as for us?

We can't fully discuss mass without units of dimension. Let's look at the elements using a mass spectrometer. The mass spectrometer allows us to determine the composition of a particular sample of material or matter. We can see the lowest element of hydrogen quite easily; it is ubiquitous throughout the known universe.

As we move up the atomic number scale, we go from gasses to solids and still are able to determine the makeup of both elements and compounds. If we can communicate about an element such as lead or gold or helium, then we perhaps can proceed with fruitful talks about mass, volume, and then length.

There is no idea here of speaking about electrons, neutrons, or other minute particles since we cannot precisely define them for our own purposes. Ongoing research continues to identify subatomic particles and their characteristics.

We still don't have a means of speaking grams or volume until our alien colleagues can identify the same elements and have a way to measure them. Once they can do that, we can then discuss the mass in grams per cubic centimeter, thereby giving us a weight and dimension for length, width, and height. (We define that the volume is in a cubic configuration for this exercise.)

How about simply just getting together and discussing the surrounding celestial bodies as to their distance from us and their apparent size? This will then allow length to be described sufficiently for correlation with our galactic friends.

We find the ability to now weigh things. Since we are out in space and not within the gravitational influence of any large massive objects, we can determine that the weights are mass and not the traditional weight associated with gravitational effects. With this, we can now define how much we have of a particular element and then what volume this amount of material must displace.

We can now define a spherical or other regular geometric shape (perhaps that of a container) that will comprise that particular amount of matter. Now we have a universal basis for discussions on dimensions and length. We can say what we call a centimeter, meter, kilometer, etc. We can understand what our colleagues use as their systems of length and thereby correlate data one to another.

Now let's go for time measurements. If we can easily discuss the elements, molecules, basic chemistry, and physics, we can then determine the radioactive decay time of certain isotopes. We can define the half-life of radioactive decay, which will be found by counting the emissions of an elemental sample versus time degradation of the quantity. Now we can speak of seconds, minutes, and years.

Yet another approach that will aid in a good first approximation of the measuring of time is the measurement of the distance of travel of a designated body in space, or watching its travel over a prescribed distance and then declaring that time as "X" time. We can only get to a working first approximation of time using a nominal distance per second of travel. Further refinement of this approach can be made later upon better definitions of measurement conditions and firmer a grasp of parametric data.

Perhaps we can agree that through much conversation we can all come up with a fundamental understanding of the way all our other colleagues measure and perceive our common immediate surroundings, and eventually we can devise a means for precise correlation. It all must begin with the understanding that time is a simplified way for us to relate to the sequential stream of events that is continually unfolding throughout the cosmos.

It is, in fact, a fabrication of a means to describe a more esoteric relationship between two or more real events and to provide a means to relate these events through temporal existence, sequential relationship, and duration with respect to others. The passage of time is a universal constant and continues without regard to our meager approach to define it. What we call a minute or hour or day has no impact on its procession. Our time measurements are tied to our perspective of the functions observed within our solar system—the earth revolving on its axis and orbital travel around our sun. None of these have any meaning to beings from another planet or galaxy.

We must take care in trying to define the correlation means of this important quantity, a correlation that is a basis for everything we attempt to measure or describe. It is, after all, a fabricated abstract means of attempting to describe the sequence of real events.

We now find the idea of communicating with our new colleagues to be less stressful and certainly a basis for good informational content being passed from one to another. Now we can look at our coordinate systems and put units of length on the axes.

At this time, let's have another colleague join us so that we now number three in total. All will have identical systems of coordinates and understanding the same basic ideas of length, mass, and time. What can we say about our coordinate system as compared to each of our other colleagues? Not much. It's possible that all of us, in just floating about, have some elements of axial rotation of our coordinate system as compared to the others. We don't have synchronicity.

While we all have the ability to use units of length, mass, and time to advantage and have those parameters identical for all systems, we must ensure that we don't have axial spin occurring between the coordinate systems.

We have chosen distant objects to tie our coordinate system to and suggested this approach to our colleagues. While it is desirable, indeed essential, to have our frames of reference be coplanar with each other, we can still drift or move about while maintaining the coplanar attitude necessary for sensible initial communications.

We can consider that the Z and Y axes are always coplanar and allow the X axis to be movable with respect to the others. Discussions along the X axis to us will be the same X axis to our colleagues, but the points of origin of these axes will translate with respect to each other. *This is relativity in its simplest form!*

Watching a celestial event now gives us the opportunity to not only observe the event and say how it looked from our perspective but to actually correlate the information so we can now define the relative relationship of our coordinate systems and thereby communicate to our other colleagues in a definitive manner.

It is important to note that each of us have our own frames of reference, and no one in our group has the preferred or prime coordinate system. Each of us is at the origin of our own respective coordinate systems; it is these coordinate systems that require further correlation in order to determine the relativity of observations between them.

Mathematical correlation done by us for the data sent by or to our colleagues will not be the same from system to system. How we see events and compare the data will require analysis as to units of length and time. Professor Einstein suggested a relativity of simultaneity, which is certainly intriguing. Lorentz suggested a transformation of the equations used to correlate the coordinate systems relative to one another. Ingenious!

FRAMES OF REFERENCE-THEIR OWN PERSPECTIVE

Let's now begin to try to define our situation with respect to the others and the universe in general. Firstly, again let's call our local region of space the local space partition (LSP). It's one wherein the measurement of single events in the immediate vicinity have identical time relationships for all who observe them, barring motion of the observers.

We are close enough that any significant distance from a celestial event is essentially identical in time of occurrence without regard to the travel time of the photons or other radiations emitted by the occurrence. All of us as a group gather in the local space partition, just a few tens-of-kilometers apart from one another.

It is essential that we keep the Y and Z fundamental axes locked together on common planes to achieve the coordinate system synchrony we require in order to have "identical" results upon observing an event and then subsequently communicating about it. Let's make the leap and consider that we've been successful in doing that, and we can now watch events and discuss them without correlation problems.

But wait! We need to ensure that there is no translation of one colleague's coordinate system as compared to the others. The Z axes must be fixed together at the point of origin, as must the X and Y axes.

L.S.P. CO-ORDINATE SYSTEM

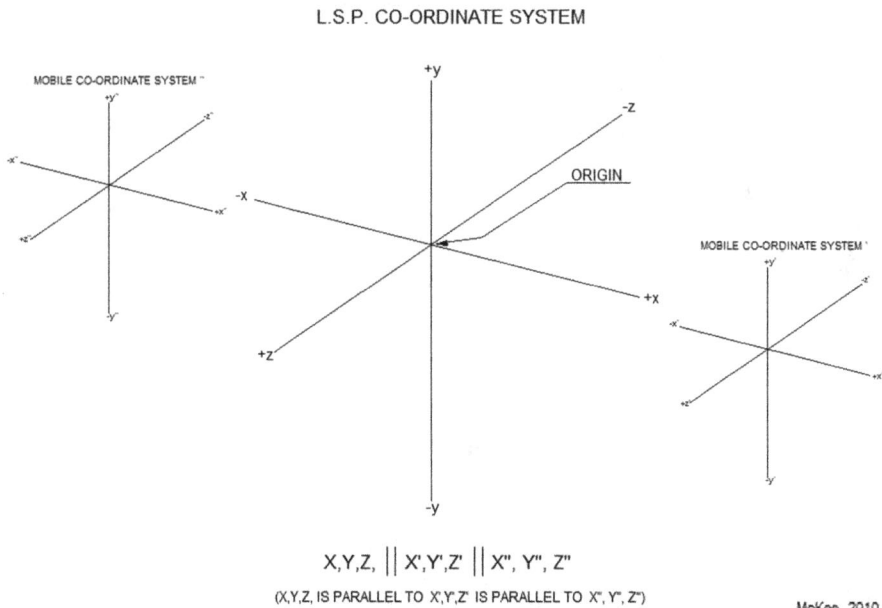

X,Y,Z, || X',Y',Z' || X", Y", Z"

(X,Y,Z, IS PARALLEL TO X',Y',Z' IS PARALLEL TO X", Y", Z")

McKee, 2010

L.S.P. COORDINATE SYSTEMS

If we are all drifting about at the same velocity (again as referred to as LSP), then we can know we have a solidly locked and synchronized group of reference frames. The answer to this is to use the LSP and its X, Y, and Z coordinate system as the master system for data gathering and then to correlate our individual systems with that one.

Now, how about time? What exactly is time? Perhaps it is simply a concept that allows us to relate occurrences in our sphere of perception to ourselves, thereby enhancing our ability to quantify and analyze. Perhaps we'll find that time is an abstract measurement of an interval of being. Until we find a means to correlate our time measurement methodology to the others and define what we mean by time, we remain wanting in our abilities to clearly associate our perception to theirs.

If we spoke with extraterrestrial beings about defining one minute of time or one second of time, they would have no way of knowing what that interval actually was with respect to them and their world. We would have to devise a message with a timing pattern and sequence characteristics that would carefully delineate it; transmission of such a signal may or may not be perturbed in its content during transmission and reception, but without successfully accomplishing that, we would have no hope of speaking "time" with them.

Now, some people will say that we can measure resonance oscillations or perhaps radioactive decay rates of elements like, say, cesium or uranium. We could thereby determine that some number of these events will constitute a unit of time passing. Some speculation in the past and up until the present day has alluded (or hypothesized) that earth time is not necessarily space time—or Saturn or Mercury or Andromeda or Galactic time!

This same argument holds for other units of measure as well. How do we relate these to our celestial colleagues? If we wish to speak of relativity with them, we must begin with a lot of preliminary preparation in word and phrase definitions not fully considered with respect to intergalactic correlation. We obviously have a need to be able to communicate these amongst ourselves as well. Until we can say for certain that all possible criteria have been considered, perhaps we should be a bit cautious in trying to definitively say what a second of time is!

We have taken physical measurements of our earth and made them the basis for initial length measurements. The earth is a planet in constant change, and in the interest of planting a firm and reasonable "stake in the ground," we must release the earth itself from this effort and be a bit more precise and repeatable in our definitions of units of length.

Perhaps we should make the comment, however pedestrian, that measurements of velocity require both distance (length) as well as time in order to produce valid results. Now we find that some individuals are desirous of changing our well-conceived units of length with contraction and our seconds by dilation! It's almost like a toddler learning to walk being bowled over after a day or two of concentrated effort and modicum of success.

It will be in our immediate interests to maintain commonly agreed upon time-and distance-measuring techniques for now while we consider other items of interest such as actual distances, actual and relative velocity, and the like. If we set about changing our fundamental tenants without an attempt at absolute and universal definition, we could easily become lost and mired down in a hopeless quagmire that would confound the brightest minds and offer only frustration to those who would seek to understand.

With respect to those who have given so much to the effort of defining fundamental parameters and units of measure, I'll proceed as if they were, in fact, not subject to relativistic changes. The outcome is interesting. When we look at the time dilation and length contraction theories that have been considered for many decades now, we find no rationale for the original hypothesis.

If we go about trying to identify why a certain experimental result was null, when the experimenter had preconceived ideas about the outcome and was shocked with his inability to show the necessary data to back up his ideas, then we all fall victim to a vain and futile effort that surely will derail and cloud our vision as we attempt to move toward reality.

There have been numerous people working on definitions, understanding and describing occurrences in our physical world, and then applying what they find to the universal realm. Using terrestrial experiences and understanding may, of course, be proper or not. If we attempt to use our ideas on mechanics, time, distance, space, velocity, and so forth, we must be quite careful that those concepts, ideas, and notions are reasonably correctly applied to the observations being made.

What would happen if we applied the time dilation and length contraction ideas in error? Perhaps the size of the visible universe would be smaller, perhaps *much* smaller than previously thought. What does this do to age considerations of the universe?

While we're at it, what exactly is a light-year, anyway? Is it the distance a photon can travel in one year's time? Really? What if photons of various wavelengths vary in their absolute velocity by some small amount? What if they encounter external forces acting upon them, such as gravity or friction, or some sort of astral-lensing that deviates their paths a bit? Some people believe that time is not a constant in the universe, which would certainly make the space-time analysis very interesting indeed!

PICK A THEORY, ANY THEORY

We have people trying to devise a space-time continuum theory, some sort of cosmic foam theory, wormhole theory, black hole theory, string theory, and the list goes on, ad nauseam. Some of them actually profess to know enough about these that they put lectures and video presentations out professing them.

Without dissenting voices, these concepts will soon become, de facto, accepted as correct, and off we'll go on another wild ride through the clouds. Who can possibly present arguments against such beautifully and professionally compiled presentations?

Beware of the salesmen in scientist's clothing! They may represent the very antithesis of reality and progress. Their polished sales pitches, thoughtful presentations, and large financial backing are immune to attacks by opposing voices that have no similar means to be heard

3

PHOTON SPEED: A CONSTANT?

Let's begin a short experiment that will open a discussion that has been waiting a long time. We'll have one colleague act as the base and serve to define his coordinate system as the stationary LSP position of origin for our first experiment. Now I'll move my craft off toward the +X axis, and the other craft moves off toward the −X direction an equal amount. We'll await a signal from the origin to start our movement up to and past the origin point.

It should go without saying that we will maintain sufficient distances from one another so no collision will be a reason for concern to our efforts. We get the start signal, and we're off. We gain speed rapidly and, as we proceed past the origin point, each of us will stay 5.0 kilometers (km) away from one another and the point of origin. As we travel toward, and then past, the point of origin, we all are observing the movements of the other spacecraft. The first craft going by is moving at 0.5 c; the second one (our craft) is moving at 1.0 c. As we converge on the origin, we will both pass the origin point at precisely the same moment. When we finish this test run, we all get together to discuss the results.

Let's pose a very basic question. If a spacecraft emitting a green light is moving with respect to a LSP, and the light being emitted is moving at c with respect to the moving craft itself, what is the velocity of the photons as seen in the LSP point of origin? Do the photons moving in the various directions look the same to an observer who is stationary in the LSP?

We have initially defined that photons are emitted at c with respect to their source. This, of course, is a specific green velocity, since there may exist a range of associated velocities, not just that single one. Photons emitted by a white light source will emit the photons in every direction in a wide variety of wavelengths. Inserting a green filter will allow that particular color to pass through and stop all others.

We will agree that the photons will be emitted in all directions equally. If we use our new speed-of-light test machine, we find that the green photons are moving at a particular velocity near 300,000 kilometers per second (km/s). This velocity is with respect to the source of emission, in our experiment, the spacecraft itself. Let's use our other mobile colleague as the subject of this thought experiment.

We also know in this case that the source is moving at 0.5 c with respect to the LSP observer. Does that mean the photons that are emitted in the same direction in which the source is moving are traveling at 1.5 c with respect to the point of origin of the LSP? Yes, indeed that is the case!

We also find that the green photon stream being emitted by that craft is seen by the LSP as being shifted spectrally toward the blue end of the photopic spectrum.

We can also conclude that the photons are moving away from the rear light of that craft as it passes the LSP point of origin. They are seen moving at 0.5 c with respect to the same stationary observer as the craft is receding away rapidly. At this point, the photon stream appears to the LSP point of origin as being red-shifted.

Important! The observer at the point of origin sees the oncoming photons as blue-shifted. The photons from the same emission source and passed through the same filter, which source is now moving in the opposite direction, would show as red-shifted. No change in the emission of the photons has been made, no change in the filter, simply a change in the velocity of the emission source with respect to the observer.

The math describing these observations can be given by:

$$v_{pa} = v_c + v_s \, sine \, \varphi$$

if the source is approaching the observer. Or by:

$$v_{pd} = v_c - v_s \, sine \, \varphi$$

if the source is moving away from the observer.

Where: v_{pa} is the resultant velocity of the emitted photons with respect to the observer for objects approaching

v_{pd} is the resultant velocity of the emitted photons with respect to the observer for objects departing

v_c is the velocity of the emitted photons with respect to the source

v_s is the velocity of the source with respect to the observer

φ is the angle of the observer, at the time of reception of the photons, with respect to the direction of travel of the source

The obvious conclusion is that blue-shifted photons and red-shifted photons are *not* moving at the same value of c with respect to the observer! The secondary conclusion, therefore, must also be that photon velocity is *not* a constant throughout the universe but varies relative to the velocity differential between the source and observer.

In the example given, it makes no difference if the source, the observer, or both are the moving entity; the result will be the same.

The colleague at the point of origin in our LSP saw the slower craft moving quickly past. That craft showed as blue-shifted light as he approached and then red-shifted as he moved away.

The second craft, moving in the opposite direction (us), showed a characteristic *highly* blue-shifted photon stream; then, as the craft went past, the photon stream disappeared! The second craft was receding from the LSP at precisely the same velocity as the photons that were being emitted toward the point of origin by his craft. His craft was moving away from the origin, so the photons had a resultant net zero velocity with respect to it. These photons don't get to the LSP point of origin. As that craft receded into space, the photons emitted by it (to its rear) were simply being laid out along the LSP at the place where they were emitted. Their velocity with respect to the LSP is *zero*! They aren't seen by anyone in the LSP as photons.

One is reminded that there is a transformation of the mathematics that can be used to explain what each has seen. According to some people, the work by Lorentz and his transforms would come to bear here. Clearly, the movement of objects in space has no affect as to the objects' physical dimensions, such as length contraction.

One is also reminded of the extensive work done by Einstein on the theory of special relativity and his train/pebble experiment. He based some of his work on this very experiment. He contended that the pebble moved in a straight line for one observer but in a parabolic arc for another.

He asserted that the pebble had no reference coordinate system of its own but was constrained by the system of the observer. It is clear in our experimental example that each of us have, indeed, moved within the LSP on a well-defined path.

Our motion has been due to forces applied by our spacecraft, not the LSP point of origin, but it resulted in a path of travel with its own set of reference parameters nonetheless.

Each of us has our own perspectives and frames of reference, and each of us is at our own respective point of origin.

Every object in the universe has its own reference coordinate system, with the object itself being the point of origin for that particular system.

Any mobile object with its particular path of movement may be seen by any number of observers from any given point of view, none of which define its path of movement, physical characteristics, or velocity. The observers are *just* observers.

The pebble dropped from the train had a well-defined path that followed the laws of physics, and no number of observers, no matter what they thought they saw from their perspectives, had any effect on the path of the pebble. The pebble was at the point of origin of and moved along with its very own coordinate system.

TRAIN & PEBBLE EXPERIMENT

In our experiment, the path as seen by the two craft going by would be different for each of them according to the application of the Lorentz equations, but the actual physical entities were not affected by the transformation of their calculations. They still had the same mass and the same path length and moved the same amount of time, unencumbered by anyone's calculations or observations external to their particular movements or local reference system.

No time dilation! No length contraction! Whatever the observers see and whatever mathematical transformations they make have no bearing on the actual path of the object being watched, nor on its length, nor the time period in which it moves through that path. Einstein erroneously impressed the results of his calculations onto the object instead of correctly impressing them solely upon the observer's data.

The time dilation and the length contraction are strictly fictional mathematical derivations. They are not physically manifested upon the object being observed! Again: no time change, no size change, and no mass change! Our recent experiment has just proven such things as erroneous and preposterous. Assigning a physical change to an object we are watching cannot change the physical character of the object or its actions, but it may very well be seen as that from the observations of others watching the same object in the same time period from different perspectives. Simple correlation mathematics suffices to ascertain the true path and time involved in the observation and to communicate those observations to others.

Let's try another example to see if we can better understand the mechanisms involved and how they actually work. We have our friend who is stationary in the LSP. We have two spacecraft that move about. Let's go out a few million meters, both our other mobile colleague and us, he in one direction and we in precisely the diametric opposite.

Now, as we begin moving toward the LSP, we are all emitting photons with our green light sources. We both move in toward the LSP at 0.5 c. Now, we look at the photons emitted by ourselves, not the other fellows, and we find that the photons leaving from our craft are moving away at the c speed of light and are green.

Our other colleagues are finding exactly the same results about their emitted photons. Now we look at one another, and we observe that the LSP light shines with a slight blue-shift of their emitted photon stream.

Our mobile colleague, in moving in our direction (at the speed of light from our perspective), shows as highly blue-shifted indeed!

Our colleague at the LSP sees both of us approaching and showing the characteristic blue shift of approaching photon streams. Our other mobile colleague sees the LSP craft as blue-shifted and us as *highly* blue-shifted. As the stationary craft and the other mobile craft again look at their own lights, they both find that they are still emitting a green light: no spectral shifts at all.

It is clear that each of us in observing the photon emission from our own crafts, and finding them moving out at c with respect to ourselves, are observing what anyone anywhere in the cosmos would observe. We, in our frame of reference, are observing light speed with respect to our frame of reference only.

Each of our colleagues observes precisely the same thing; therefore it is impossible for c to be a constant in the universe! Our observation of the speed of the photons emitted by our rapidly approaching colleague must include the relative velocity of his craft with respect to ours as well.

If we release the idea of the stationary LSP for a moment, we find that each of us could be considered either stationary or mobile and still observe precisely the same results.

If one wished to use his particular frame of reference as the prime frame of reference for a particular occurrence, any discussions with others in a different frame of reference might need to include the mathematics to correlate such an event and observations into the other's frame of reference in order to make sense of it and accurately communicate.

For this purpose, the mathematical approach taken by Lorentz seemed reasonable, but it should be avoided as false. The mathematics of intersystem communications notwithstanding, it is important to understand and state the universal requirements of the physical characteristics of photons and their discharge from photo-emissive sources.

Let's get back to the spacecraft and continue our journey. Say that we stop for lunch somewhere, and another craft exactly like ours parks alongside. When we get ready to leave, the other craft gets out of the parking lot first and is coincidentally going our way.

As both crafts accelerate to the speed of light and then beyond, our new friend will still see us in his rearview mirror. What will we see? Well, we know that both craft have a small green light on the front and rear; his light will have been seen by us as green and remain so.

Now let's look at this scenario as Einstein would have seen it, he saw the velocity of the emitted photons as being independent of the velocity of a moving emitting source.

The fellow driving his spacecraft ahead of us and speedily going along will no longer see our light in his rearview mirror. If he were watching closely, he would have seen the light become red-shifted and then disappear beyond the capability of his eyes to see. We would see the light on his craft begin to move toward the blue end of the spectrum and then into the ultraviolet portion of the spectrum and disappear beyond the capability of our eyes to detect it.

Let's get this right for a final time. If a light source is stationary in a particular local space partition, the photons are being emitted at a velocity that is proportional to the energy level of the photon beam observed. The human eye will see this as a color. The complex color band consists of a broad range of photon energies. A "fingerprint" of such spectral spreads is commonly observed, as they are for distant galaxies; the only difference is that the galaxies show us elemental content by way of their fingerprints.

There are really no surprises here. Scientists have seen this spectral shift characteristic of astronomical bodies-in-motion for years now. The spectral fingerprint of an observed galaxy will lay out the spectral difference as the observer looks first at the front-on view of the galaxy as the component stars are moving in a sideways movement with respect to the observer, then receding (red-shifted), and then approaching (blue-shifted).

Another caveat is that using c as the speed of light is erroneous and only approximates the true character of a photon beam. Close considerations of the nominal c value will demonstrate that there is an energy spread from high levels seen at violet and beyond to low levels at deep red and beyond. Without doubt, the spectrum of the photon beam extends far beyond these simple limitations from a kinetic energy level of zero, with respect to its local space partition, to some yet undefined upper limit beyond the presently prescribed relativistic limits.

There is no logical reason to consider the velocity of photons to be either fixed, constant, or of a limiting nature with respect to an absolute maximum velocity achievable by other objects in space.

Clearly, we have mixed apples and oranges in our efforts to understand and identify the similarity of actions and reactions between electromagnetic and particulate flow. While this error is easily made, we must begin to differentiate the radiations analyzed not only by actions but by the sources from which they originate; clearly they are not all the same, however similar their physical characteristics appear to be.

Now we have this fellow in his spacecraft speeding along with his green lights emitting brightly. The idea is that this fellow sees his light as green. No matter how fast he goes, with respect to some other frame of reference, his light stays green to his vision.

His photon emissions are still the same in his frame of reference. They are only different to observers in different frames of reference. The photons don't know they're traveling much faster as far as the stationary observers are concerned, and they don't have an oscillation or frequency associated with them that changes as they travel in a particular direction.

We will have to be careful from now on in defining the spectral character of the lights we use or observe since we may have significantly different velocities between the emission sources and ourselves to deal with.

We will need to call out a particular frame of reference for our green" light and perhaps even consider defining the specific wavelength characteristics of the content to ensure that we can predict the observable shifts and have the ability to measure them. I have included a tabulation of expected results as a result of travel speeds and direction.

RELATIVE "SPEED" (RELATIVE TO "C")	"WAVELENGTH" RECEDING 552 NANOMETERS -STATIC	"WAVELENGTH" APPROACHING 553 NANOMETERS -STATIC
"C" +/- ↓		
0.005	552.75	547.25
0.01	555.5	544.5
0.02	561	539
0.04	572	528
0.06	583	517
0.08	594	506
0.1	605	495
0.2	660	440
0.3	715	385
0.4	770	330
0.5	825	275
0.6	880	220
0.7	935	165
0.8	990	110
0.9	1045	55
1	1100	0

DOPPLER CALCULATIONS TABLE

Now we'll use Euclidian-Newtonian space and coordinate systems, photons with emission velocities linked to and dependent upon the source velocity. As the craft passes the speed of light velocity, the photons emitted by its little green light will have a velocity of zero with respect to the LSP. The other craft, as well as our own, will be moving at c with respect to the local space partition.

A local observer would be able to detect us approaching with light shifted toward the blue, and as we pass, the light would flash toward the red and disappear. Assuming their eyes are limited as are ours, they will not be able to *see* the photons but can only detect them as particulate energy levels.

4

CAPTURING PHOTONS

Let's look at the photon for a few moments. If an engineer or scientist wants to look at light in general, perhaps to measure the amount present at a particular place and time, he would commonly use a photometer. Numerous photometers use very sensitive photomultiplier tubes with photosensitive photo cathodes that capture the photons and subsequently convert this information into the release of photo-electrons internally.

The electrons are typically electrically accelerated toward a metallic electrode within the tube and, upon impacting they knock a greater number of electrons free from that electrode surface. These "secondary" electrons are likewise accelerated toward yet another electrode, whereupon even more electrons are knocked off of that surface. This process, known as electron multiplication, continues through the electrode array and finally to a collector that produces a measurable electrical output from the tube that is proportional to the amount of light captured at the photocathode.

Focusing
Cathode   Electrode   Anode

Photons

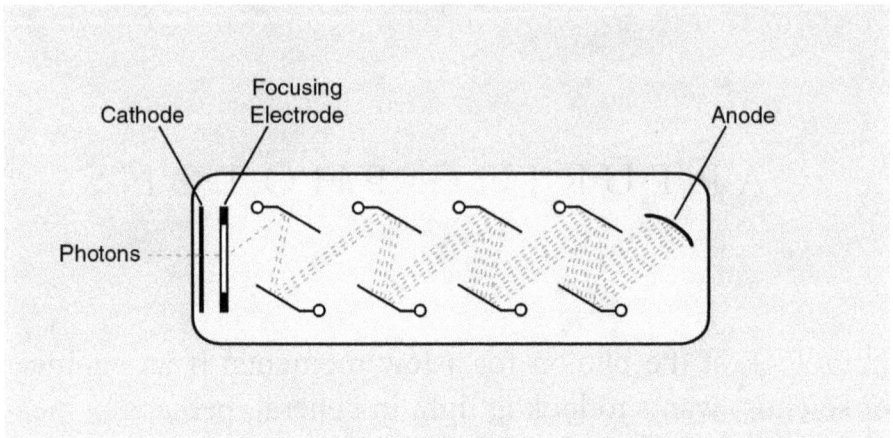

PHOTOMULTIPLIER TUBE

If ultrasensitive photomultiplier tubes are used, they are often cryogenically cooled to reduce spurious noise caused by voluntary electron emissions within the photomultiplier's internal components as well as amplifier noise involved in the measuring process. This spurious "background" noise is typically caused by subatomic activity due to the ambient temperature of the device; it is claimed that by bringing the apparatus down to an extremely low temperature, very low numbers of photons can be detected.

Some scientists claim to be able to measure single capture events with an even more sophisticated apparatus. (Takeuchi, J. Kim, Y. Yamamoto, and H. Hogue. "Development of a high quantum-efficiency single-photon counting system," *Appl. Phys. Lett.* **74**, 1063).

Some great strides have recently taken place using Silicon Avalanche photo-diodes (www.stanford.edu/~sanaka). Strides in various laboratories throughout the world and with a number of electronic component manufacturing corporations have shown much success.

When the solid-state devices detect incoming photons, they react by exciting electrons in the very sensitive impurity band of such devices. Since thermal contributions to noise obscure the sparse number of electrons being excited by the incoming photons, cryogenic cooling is still in order.

There is typically a chart provided with the photomultipliers that map the photocathode sensitivity with respect to incoming color described in terms of wavelength of light. This same type of sensitivity chart is also applied to solid-state photosensitive devices.

The solid-state devices are used for applications where less sensitivity to incoming photon flux can be tolerated. The solid-state devices can be very small physically and thereby very useful in many applications where size of components is at a premium.

We have assigned the colors of our own physiological characteristics to our instrumentation used to detect photons. We have found that capturing them can be done and if we assign the resultant device data with chroma information commensurate with our experience with our own eyes, we can comfortably correlate the received data. This doesn't make the color assignment real for the photons since they don't change; the assignment is only for us.

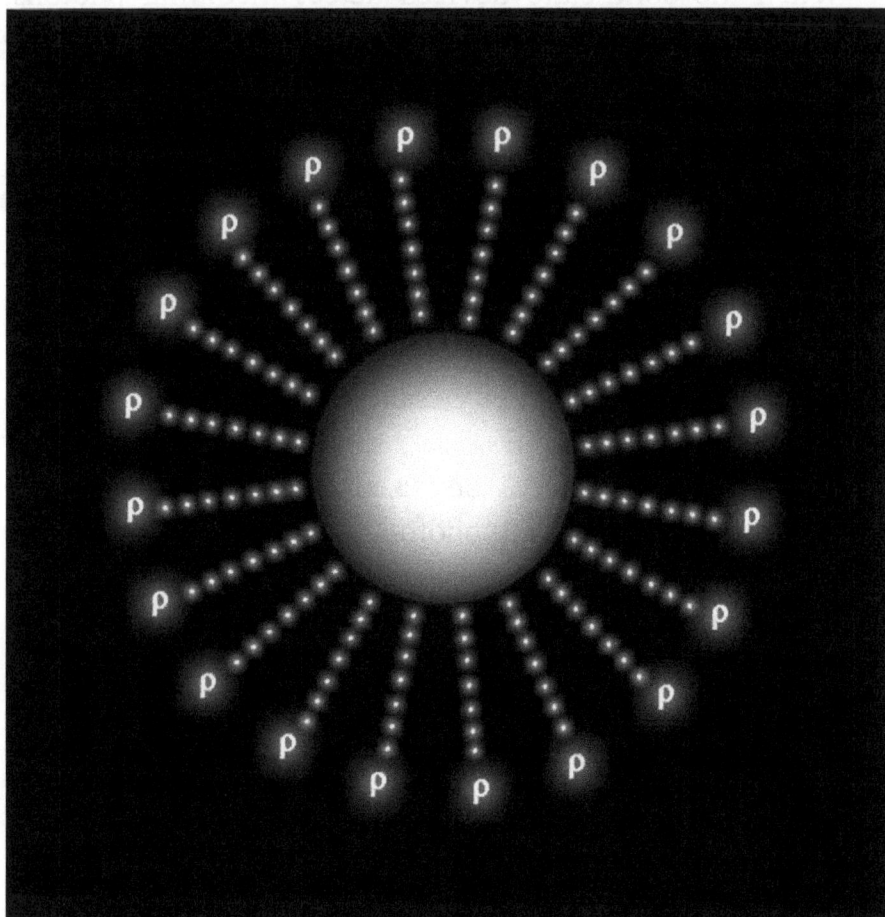

PHOTONS EMITTED IN EVERY DIRECTION

We know that the photons emitted by the source move in all unconstrained directions and then only are constrained by the physical characteristics of the source and objects in their paths. If the source's entire surface is emitting photons, then they are moving out in every direction at once. Now let the source move away from us at a near relativistic velocity, and observe the red-shift phenomenon.

Let the source be moving toward us in a similar manner, and observe the blue-shift. These shifts are compared to the spectral character of the emitting source if such source were to be moving tangentially to our position, that is, across space without any appreciable advancing or receding velocity.

The photons have not changed. They *all* are always being emitted with the same characteristic emission defined by that source at that specific temperature. The only thing changing is the differential velocity of the source with respect to us, the observers.

We are now required to rewrite the idea of the Doppler effect since we don't have a wavelength to work with and the speed of light is not a constant; it changes as the difference in velocities between the source and detector change, becoming a critical factor.

If we look at the original Doppler equations as applied to the Michelson-Morley experiments, we have:

$$v_{in} = v_{out} \sqrt{\frac{1 - w/c}{1 + w/c}}$$

Now we must derive our equation to utilize the known factors without regard to the fixed speed of light typically used.

If we accept the premise that the photons can travel between zero and some velocity at which they are emitted from their source, we can then go about a derivation between these limits.

If we use an approximate center of the photonic spectral range and move outward from there, we can watch the wavelength change rapidly. We will go from zero relative photon velocity through green light at 550 nanometers and then an equivalent velocity above green as a reasonable range for our mathematical derivation. Soon, we'll see that we have put ourselves into a mathematical corner, so to speak.

The results are given:

$$\lambda_o = \lambda_s \; (1\text{-}d)$$

for a receding relative velocity, where d is some fractional part of λ_s and:

$$\lambda_o = \lambda_s \; (1\text{+}h)$$

for an approaching relative velocity, where h is some fractional part of λ_s.

Everything works well until we realize that we don't know the upper limit on the velocity of photons! We can work about the photopic center of the human visual acuity, but that in no way identifies the upper velocity for photons as quanta.

We don't subscribe to the premise that photons are a part of the electromagnetic spectrum and therefore don't have a frequency or wavelength, other than to relate the calculations to familiar units of measurement and ranges.

UH-OH, MATH!

We know that the emission of photons (quanta) from a solar orb has a broad energy range and a portion of that energy range is equivalent to the human photopic region; to that, we have seen fit to assign wavelengths and frequencies. To the energy ranges beyond the visible and the near visible, we are limited to then define the range of possible photon velocities (relative to some LSP) from zero to some heretofore undefined limit.

How would you like to define the limit of energy that a solar orb imparts to a photon? Without some idea of this limit, we have no possibility of defining the resultant velocity (read: wavelength… frequency). We are, therefore, without the possibility of placing an upper limit mathematically and must simply work within boundaries left to us humans and our limited resources for discovery.

5

A MASSIVE CONSIDERATION

We know from extensive experimentation and observations over the years that for an object to have a characteristic energy level, it must have a mass component as well. Photons certainly have energy and, therefore, *must have mass,* albeit uniquely small.

Now some would have us believe accelerating objects of all sorts including photons are growing more and more massive with an increase in their velocity.

PHOTONS GETTING LARGER

Some people would even have us believe that they flatten out in the direction of that travel. Some would have us believe that the energy and mass are interchangeable quantities.

I expect you'll find some who think they're small cubical bits of matter, or perhaps foam or string-like in appearance. There are no mechanisms in physics that would sustain the idea that the mass of a solid object or its physical size and shape would change due to increasing velocity.

Now we come across a real dilly of a problem. If these particles are released from a body due to an explosion, perhaps a supernova or the like, since they are particles with a mass content, they would have to be included in determining the original overall mass of the body that exploded if one wished to accurately estimate the original mass of the object.

I can't imagine being able to count, or even calculate, such a number and, indeed, in such numbers that particles with even the smallest concept of mass would significantly contribute to the source body's original mass.

Consider occurrences as simple as burning fuel, perhaps a log in the home hearth: while the solid remnants, such as soot or ash, and the radiant heat energy (more photons?) and gasses released account for the overwhelming portion of the mass of the original log, the visible photons must also be accounted for in order to determine the complete original mass of the source prior to its burning.

Another way to look at the massive nature of the photon beam: if we were to travel along the same direction as a beam of light, we would have the ability to detect not only the energy component of such a beam but also observe the finite, albeit miniscule, mass of the photons contained therein.

We would find, as we accelerated up to the same speed at which these photons were traveling, they would be rendered to our observation simply as subatomic particles of mass moving rapidly through space. Their velocity with respect to us would be zero, and therefore they would no longer be perceived as photons but simply as minute particles without apparent relative energy, frequency, or wavelength.

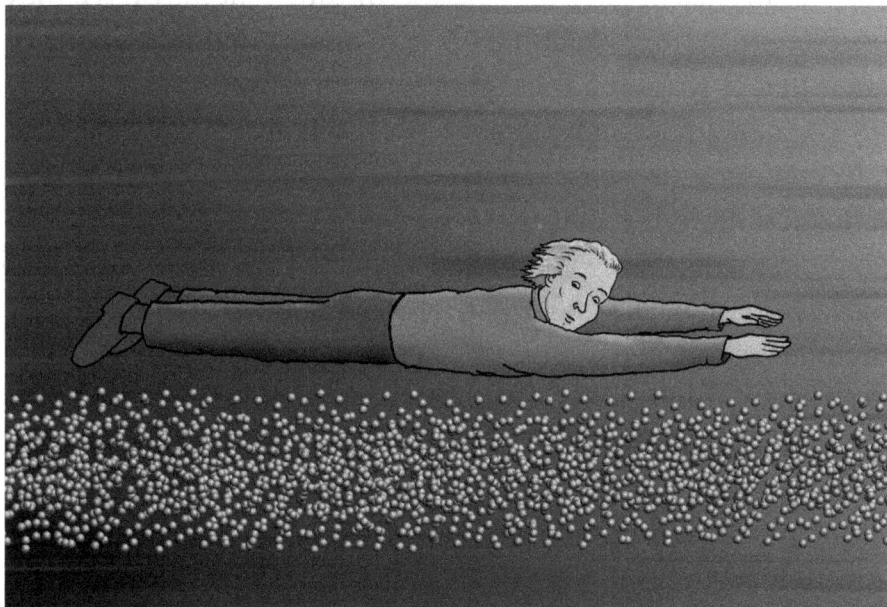

TRAVELING AT THE SPEED OF PHOTONS

Let's look at the photons emitted by any source moving at precisely the same velocity but in the opposite direction as the photons being emitted. They would be seen, if there were a means to see them, by an observer in the local space partition as just a trail of small particles of mass. If the photons emitted rearward by the traveling craft are equal in velocity of the craft itself, then they are released without velocity or direction of their own (remember, this is as seen in the local space partition only).

Unfortunately, they will be so small as to be virtually invisible to that observer; only extreme volumes of such small particles would constitute a group sizable enough to casually visualize.

If we now consider the extreme number of emissive sources throughout the universe that have been, as well as still are, contributing to the total number of such particles, and for such an immeasurably long period of time that they have they been contributing, we come up with an astounding volume of mass that is scattered throughout the universe.

We can understand that in such quantities, this particulate content in space constitutes certain and substantial significance. Imagine that these particles thus present throughout space and, being interacted upon by the huge variety of massive bodies and other particles, over time might coalesce together in loosely and clearly defined groups.

Perhaps there will be apparent voids wherein these particles will have been removed by such interactions as well. Imagine all of the suns in all of the galaxies, moving about in every possible direction and at a vast variation of velocities, emitting particles of vast quantities for billions of years in all directions, and perhaps what we end up with becomes *dark matter*!

Since these photons have mass and can be gathered together in loosely formed clouds in sufficient quantities, they could have a level of opacity to other radiations, those either particulate or electromagnetic in nature. Certainly they would manifest themselves as something other than what we have come to think of as photons since they are no longer photon-like in appearance in the classical sense.

Since this dark-matter concept is conceivable, and since photons have been emitted, reflected, and scattered by numerous sources in various states of motion, it is also conceivable that over time this dark matter has been built up to a significant level here and there throughout the vastness of our universe and will continue to build up throughout time. A quantity of such matter will undoubtedly be used in formation of other matter as well as various astronomical bodies and suns over the ages.

We can imagine that there will be areas in which the concentration of this dark matter content varies so that some regions will be essentially clear of such flotsam and others will be virtually opaque.

Now we can imagine that over eons, these particles will have had physical, chemical, and atomic interactions and thereby coalesced into larger particulates, clouds, and food for the numerous "star factories," a recycling process in its purest sense. Perhaps some of the beautiful structure and range of textures found in Nebulae are prime examples of this process.

Since these photons and other such components have mass, this also means that they could be (and indeed are) affected in their flight paths as they pass by massive bodies such as planets, large asteroids, and the like.

If we can mentally conceive of a subatomic particle that has mass, as we know they do, we can then move along the logical path and understand the possibilities of photon velocity versus path deviation due to gravitational pull (read: mutual mass attraction). We can also understand that the deviation has a small inverse proportionality to the energy level (read: specific relative velocity) of the particles in question.

Without question, now we must also conclude that astronomical turbulence will also prove influential in the photon's movements as it travels through the cosmos. One can imagine the coming together of solar winds from all over: every solar system, galaxy, nebula, and rogue star's contribution into a volatile turbulent soup that causes, in simplistic terms, what is allowed by some as space-time distortion.

Now it has been found through close observation by clever scientists that such a deviation in flight trajectory has been found experimentally on a number of occasions and that deviation cannot be explained by refraction due to interaction with a planetary atmosphere.

Many massive bodies that, in fact, cause such a deviation do not have an appreciable atmosphere sufficient to cause any refraction. These results have been well documented. This is wonderful; we're making progress now!

Some would have us believe that there is a warping of space-time and such warping controls the trajectory of particles and massive bodies in their journeys through space.

Perhaps what is seen is simply the interaction due to mutual attraction and interaction of massive bodies (called gravitational attraction), along with the varied turbulence that might be encountered due to a mixture of myriad solar winds. Perhaps electrostatics also plays a part? Maybe some frictional components or collisions with dark matter contribute to the deviations? Perhaps particulate and molecular interactions and optical refractions or reflections participate as well?

With their small size and variations in energy level, these particles will certainly be affected by the electrical nature of molecules and atoms dependent on their own specific energy level and physical composition.

This would lead us to assume that such interactions would form the basis for refraction as well as reflection of such particles.

One is reminded here of the dispersion characteristic of prisms wherein the light beam approaching the side of the prism encounters the prism surface. Could it be that the energetic photons simply are subjected to a vectored reaction within the glass-specific matrix? Perhaps similar to energetic electrons in a magnetic and electrostatic field, to be deviated from their path, they are providing a spectral dispersion effect? Do you suppose that classical glass optical physics is based on such interaction between these glass molecules, surfaces, and the individual photons?

It is certainly conceivable that they may even demonstrate gross diffraction patterns on a massive scale (read: astronomical). There will be occasions where a multiplicity of images from a single source could appear to look like twins or perhaps triplets of the single source.

How about the way in which photons interact with glass and other refracting media? Perhaps we have a simple (or not so simple) chain of interaction between the high velocity photon and the glass molecules making up the matrix of a particular glass.

The resultant interaction would be a function of the physical matrices of particular glasses as would be the deviation of the photon beam impinging upon it. Perhaps we have a mechanism of collisions between the photons and the surface of the matrices such that the deviation is proportional to the energy level of the oncoming photons.

If, now, this initial contact with the matrix served to define a regularly ordered path within a glass substrate determined by the initial angle of incidence, the path direction and transit speed would then be preordered by this angle.

With the incoming photon energy spread throughout the photopic range, the resultant spread of photons internal to the glass would form a spectral fan, which upon exiting would reconstitute into a singular beam at the output resuming the original trajectory of the incoming beam (this is assuming parallel planar surfaces of the glass substrate, i.e., the same thickness overall). If the glass substrate is a prism, the fan shape may be exacerbated into a wider fan as it exits, thereby making the observation of the visible spectrum dispersion easily accomplished.

What if the photons actually maintained their velocity through such a matrix but appeared to be slower due to the vast number of course changes they would be subjected to in the transition? The velocity not actually changing and then as they leave the glass they again appear to gain speed without external forces being applied! Now really, how would they speed up upon exiting if there were no external forces acting on them?

One is also reminded of the flight paths and characteristics of electron and particle ballistics with respect to electrical fields and electrostatic lenses and magnetic fields ($\upsilon \times \beta$ vectors).

Although electrostatic interaction is unlikely in photon ballistics, particle-to-particle as well as particle-to-local-field interactions play a significant role at the quantum particle level.

A person familiar with past work done in electron ballistics using both electromagnetic coils for focusing and electrostatic potentials for accelerating, and lens effects, understands that diffraction patterns and wavelike addition and subtraction of groups of electrons is found as a routine observation. Particles in vast groups can certainly act like and can be seen to behave in a wavelike manner.

This by itself does not fully describe the actions of photons and their interactions in forming wavelike path appearances but certainly gives some food for thought if considered a contribution to our overall understanding.

6

PHOTON ENERGY?

When physicists consider the idea of energy of a body or particle, they invariably must also consider the mass of that item. The classic energy equations of Newton and Einstein spoke clearly of the energy levels being proportional to both the mass and the velocity (speed) of massive objects. Einstein took this energy relationship into the realm of the photon.

Max Planck, Niels Bohr, and Wilhelm Wein, among others, considered the thermal and subatomic involvement as well. There are those who believe that some subatomic particles exist without mass at all! Trying to fit them into such a mathematical model would be indeed difficult to do.

The point of this book is to address photons specifically, so let's consider them. The red and blue shifts that are observed when the sources and sensors are moving with respect to one another are a function of the relative changes in their associated kinetic energy due to the relative velocity of the photon mass; consider the

$e = \frac{1}{2} m v^2$ equation as Newton proposed, or Einstein with $e = mc^2$. There have been others as well.

Now remember from earlier, in our little space experiment, we had an observer in our local space partition watching as the small spacecraft approached rapidly and then passed the observer's position. As the craft is approaching, the observer will see the color of the green light as being blue-shifted. A corresponding energy level reading would show a higher energy level for the oncoming photon stream. As the craft passes and speedily moves away, the color will transition to a red-shifted appearance, and the energy level of the photon beam will be seen as lower than when it was advancing and blue-shifted.

Laboratory experiments done here on earth have consistently and definitively demonstrated the color-to-energy-level relationship of light. Consider the "black body" work done by Wien, Planck, Boltzmann, and others. Unfortunately, much work has been carried out in linking these radiation levels to electromagnetism without regard to the possibility that they are subtly different. Particulate radiation can appear to be electromagnetic in action but must be separated from it in the final analysis.

Photons that are emitted from a light source are emitted at an associated energy level. They travel at a velocity prescribing that energy level. They do not have a color or a wavelength, nor do they oscillate at some frequency. They are simply a steady stream of emitted photons having mass, velocity, and direction. In calculations involving photons, the idea of wavelength and frequency of light beams would more correctly use energy levels for precisely comparable results.

Your attention is called to the idea that the photo-cathodes and the solid-state devices we've already discussed are actually reacting to the photon's energy and not the wavelength. As we demonstrated before, the photons have no idea that they are being emitted either in a particular direction by the emissive source nor with any particular color. They obtain their spectral characteristics from the emitting body, as the energy level at which the particular photons are emitted and from the emissive body's velocity with respect to any particular observer, sensor, or detector.

For light-sensitive sensors, we will invariably find an accompanying chart outlining the sensor's sensitivity (output) versus wavelength since the sensor responds to the different energy levels of the oncoming photons. Volumes of data have been collected demonstrating that photons in the photonic region of the photon spectrum have progressively higher energy levels as the color moves from infrared to ultraviolet.

As we have established, the blue-shifted photons are moving at a higher velocity than those that are relatively stationary to the observer's position. This higher velocity with a fixed mass of the photon particle will clearly demonstrate this same higher energy result.

Look to the work done by Planck and the energy levels emitted by black bodies at particular temperatures. The idea is clear; a range of photon energies is emitted (again, in every direction), and in this particular case the energy levels are clearly associated with the thermal status of the emitting body.

Perhaps we can relate this data loosely with the emission of photons and their energy (spectral) spread from the myriad suns throughout the universe.

But wait, something had to change in order for us to perceive a spectral shift! Yes, correct! The energy levels of the photons have changed due to the difference in the velocity at which they are being emitted and subsequently detected. Look back at the work by Newton and Einstein with respect to kinetic energy. Remember, they said that the kinetic energy of a particle is proportional to the mass times the velocity squared (c=the approximate velocity of light for this purpose).

The photon mass is a constant, and the energy is proportional to the square of its velocity relative to the observer. They were wrong (albeit close) as to the precise form of their equations. By example of the previous experimental results, we must use a velocity differential between the source and the sensor along with a proportionality factor yet to be discussed.

It's time for a little fun.

Let's say that one of our cosmic group is now several light-years away in the +X direction, and another is the same distance away in the −X direction. We'll have one of our craft near the center of the LSP (us) be designated as a traveler as well. Let's consider the following simple experiment.

Our +X colleague now moves towards the LSP origin at 0.7 times the speed of light. Our −X colleague also moves towards the LSP origin at 0.7 times the speed of light. Our local traveler (us) moves along the −X axis at 1.1 times the speed of light. All our crafts are fitted with lights that emit light that appears green to ourselves. Now, what do we all observe in this situation?

We have one observer remaining stationary in the LSP origin that sees the outer fellows coming in with their lights being highly blue-shifted. He can no longer see the local traveler (us) moving away at 1.1 times the speed of light because any photons emitted by his craft are themselves moving away from the point of origin at 0.1 times the speed of light.

The traveler coming in from the +X axis direction can see the local traveler leaving towards the − X axis with his light red-shifted. He can see his colleague coming in from the −X direction with his light *highly* blue-shifted. He can see the local stationary craft sitting there awaiting his arrival and its light showing as blue-shifted.

We should understand that our light emission, while appearing green to our eyes, is not necessarily seen by others the same way. The emissive spectrum is defined by the source characteristics such as temperature, composition, and velocity as well. What happens when the travelers are moving about in the local stationary group? All see their emitted photons as moving away from their own craft at the light speed associated with a green photon. The spectral shifts that are observed in the previous experiment are strictly indicative of relative velocity between the emissive source and the observing detector.

IT'S ALL RELATIVE

Without any relative motion, all the light would still appear to all observers as nominally green. This would be demonstrable as well in having an observer on the LSP watching a group of craft go racing by at 0.7+ times the speed of light and seeing their photons go from being blue-shifted as they approach and then red-shifted as they move into the distance.

The light sources on the craft are not changing their output emissions. They still appear as green to all those who travel along with them in their particular frames of reference. They all see their own lights as green throughout the experiment. The incoming group of travelers would see the stationary fellow as he sees them, with the light showing as blue-shifted as they approach his position and then as red-shifted as they move off into the distance.

The significant conclusion to this experimental approach is that the emitted photons move at their particular speed of light relative to the emissive source. If the source is moving away or toward the observer, the observer will see the photons as moving at the speed of light changed by the appropriate color shift associated with the emitting source-to-observer (relative) velocity.

We conclude and therefore say:

McKee's Law states the speed of light, typically denoted by c, is not a constant in the universe but only within the specific frame of reference of its emissive source!

Obviously, this is not a law but is intended as a bit of sarcastic humor. The idea that we, as observers, can formulate a law that defines how an observed object behaves is ludicrous at best. We simply observe and try to hypothesize why we are observing what we are observing. The objects are simply following the laws of nature (or physics, if you must), as are we all.

The results of this experimental effort are multiple and of significant importance throughout the physics community:

A. The emitted photons don't know in which direction they are being emitted; they simply move away from their emissive source at a velocity proportional to the energy level at which they are emitted and are dependent on the character of the source. All photons are not created equal with respect to emission velocity.

B. The preceding statement bring about another thought. The emitted photons do not have a wavelength, an oscillation, or a color. They simply have a mass, velocity and thereby a kinetic energy level and finally a velocity relative to the observer or sensor.

C. A sufficient number of photons moving in concert, like molecules of water, may behave in a wavelike fashion with substantially wavelike properties.

D. Ancient photon matter from trillions upon trillions of light sources moving about the cosmos over countless eons of time have built up and significantly contributed to the apparent total mass of regions of the cosmos, such as nebulae and galaxies, and throughout the vastness of "empty" space, and have contributed to the conglomerate we call dark matter.

E. This dark matter is essentially stationary in the respective LSP where it is situated. It is moving as its surrounding cosmic material is moving and thus loosely contributes to the apparent mass of its surround. Since this material doesn't emit photons but is in fact clouds of dark and micro-mass material that at one time was photons, it has the ability to partially obscure the passing of photons that are moving through it.

F. It is clear that photons only have an *apparent* luminance when traveling at relativistic velocities. When we measure the spectrum, we are simply measuring the energy contents of the photon stream with respect to our frame of reference. The luminance is a function of its interaction with the biologic receptor, the eye.

There was once an interesting scientist who had the thought, "I wonder what it would be like to travel alongside a beam of light." This is a paraphrasing but is the essence of the idea.

Well, perhaps we can now answer that. If one is standing beside a photon source and begins to travel along a path of the emitted photons, as he gets up to the velocity of the photons, he sees that they are simply very small particles of mass. Photons moving at the same speed as the observer are shown simply to be particles of mass on an extremely small scale.

Photons emitted in such numbers as may casually be seen in cosmic events such as supernovae certainly have contributed significantly to the original mass of the source in that particular event and also now contribute to the mass in the surrounding space. The idea of such a mass content certainly can't be a new one, but it is one that would contribute to the idea that photon streams would be affected by gravitational forces near massive celestial objects.

One can also easily surmise that they are devoid of electrical charge since they show no apparent course deviation in the presence of high electrical fields. If the polarity or atomic electrical charge is dependent on whether or not an electron is present or missing, then the idea that the photon is so small that no electron (and therefore no atomic charge) can be associated with such a minute particle renders it neutral with respect to an electrical charge. One wonders, if a photon has no electrical charge, can it truly be considered a part of the electromagnetic family of radiation?

Let's consider for a moment what photons become once they move out of the bands identified as infrared through ultraviolet light. If they are moving slower than the far-infrared, what do we call them?

If we look at the typically accepted electromagnetic mapping chart we find that micro "waves" and radio "waves" are found down there. If we go above the ultraviolet, we find X-rays and gamma "rays." It's curious; we go from radio waves through microwaves, photopic or visible light, and then through X-rays and gamma rays. From *waves* to *light* to *rays*!

Well, what are they? Are they really waves and rays? Do photons (light) turn into other types of radiation, cataloged as electromagnetic, or are they really simply particles?

Perhaps we should look at some salient characteristics of the various identified "radiations." Radio and micro electro-magnetic waves are thought to have electrical and magnetic fields as component parts; dependent on frequency, they travel along in different modes such as line-of-sight or perhaps hugging the earth's curvature. They demonstrate a characteristic possession of energy without physical mass.

Gamma and X-rays, also cataloged as electromagnetic, must also have electrical and magnetic components associated with them. Depending on the frequency, we find the ability for these to penetrate solid matter. Perhaps their extraordinary velocity just moves with ease through many types of matter they encounter.

Do you suppose that all of these radiations travel at the same velocity in space? Let's look at one way used to generate X-rays. If we bombard a metallic target such as tungsten with very fast-moving electrons, we can easily generate X-rays. We find that the electrical field potential used to send the electrons to the tungsten target reasonably defines the character of the X-ray beam thus generated. The number of electrons in the beam, the energy level on the electrons at impact with the target, and then any coincident filtering through interposing material will serve to clearly define what the X-ray beam will look like.

Now the efficiency of conversion from electrons striking the target to the generation of X-rays is not a 100% conversion, so we end up with heat generated in the target. We also have a distribution of output X-ray energies known as bremstrallung. This embodies the energy level of the impinging electron beam, the material type used for the target, and again subsequent filtering of the X-ray beam.

What do we get if the energy potential that drives the electrons into the target is continually decreased until it goes to zero? Well, we can understand that the quality of the X-ray beam will continually drop proportionally with the accelerating potential until finally no X-rays are emitted.

Understand that during that entire process we are generating *heat* in the bombarded target material. We continue to generate heat at lower and lower levels until the target is no longer bombarded by the electron beam.

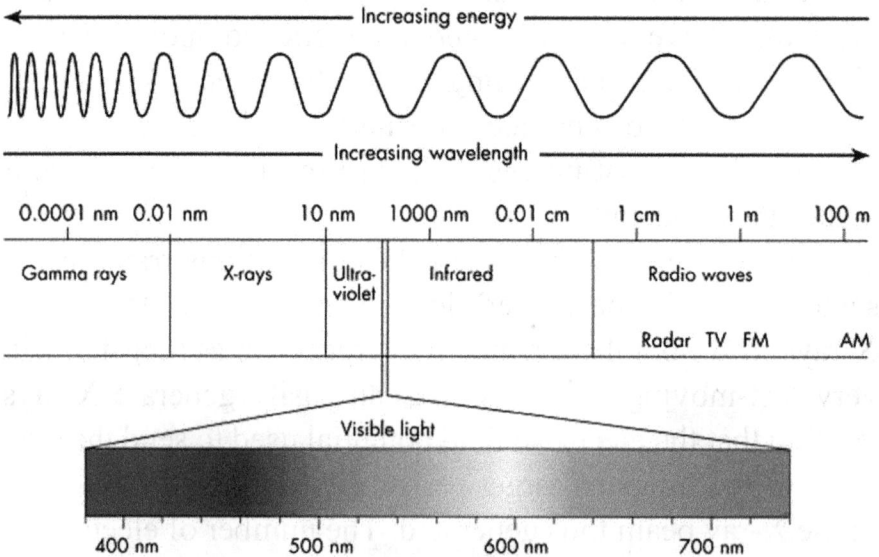

◄──────────────── Increasing energy ────────────────

Increasing wavelength ──────────────►

| 0.0001 nm | 0.01 nm | | 10 nm | 1000 nm | 0.01 cm | 1 cm | 1 m | 100 m |

| Gamma rays | | X-rays | Ultra-violet | Infrared | | Radio waves | |

Radar TV FM   AM

Visible light

| 400 nm | 500 nm | 600 nm | 700 nm |

"ELECTROMAGNETIC SPECTRUM"

Now let's get this right, shall we? We use lots of very high-energy electrons and get very "hard" X-rays and very hot targets, which, if hot enough, will emit photons in the form of visible light. The visible light will, if hot enough, show as "white hot" and include the entire broad photopic spectrum. As the electrical potential is lowered, the X-rays soften. If the quantity of electrons is lowered slowly, the photons decrease in intensity and energy level until the temperature reaches the ambient room temperature. We understand that there is still minimal emission at room temperature.

Other mechanisms to generate X-rays certainly exist throughout the cosmos, as is readily apparent. Without doubt, they all involve emission of photons as well. If the photon spectrum is tied to its energy level and the X-ray emission to its energy level, it is clear that they both will continue in upward energy levels.

Another significant difference between the X-ray and photon characters is that photons don't move through optically opaque materials. X-rays move through many optically opaque materials with comparative ease. One must be advised that the condition of being opaque is relative to a particular radiant beam or radiation, whether particulate or electromagnetic in nature.

The X-ray beam moves through materials of lesser mass-density more easily than through the denser materials. The photons halt at virtually any density if the object is opaque to the photons but can traverse a very dense ("Z" cross-section), optically clear material of great cross-sectional thickness with ease, some of which would stop X-rays in their tracks. The characteristic reactions of these two different radiations prove them to be distinctly different from one another.

Can we then conclude that the X-rays are not a photon emission? Without doubt! We can consider the high-energy X-ray characteristics as fundamentally comparable to gamma radiation, where we find similar abilities to penetrate items of mass. Neither seems sensitive to fields of high electrical potential. They don't become deviated in their paths of travel by high magnetic fields. Many scientists will identify the radiation source as the defining parameter for the type of radiation observed. It's time to judge the type of radiation by its characteristics of behavior as well; reference to the source characteristics and structure should be made carefully and then its velocity component relative to the observer or detector.

We've seen that high-energy particle bombardment can generate heat as well as both visible and X-ray radiation; clearly the source characteristics and mechanism for generating the radiation is not definitive.

Let's consider the photon spectral character once more. Consider for a moment a spiraling galaxy that is close enough to us to appear very large that is revolving at a furious pace. The galaxy as a whole is moving away from us.

The spectral band demonstrates the distinctive red-shift for the stars within the galaxy that are moving away from us at a greater velocity (if they are on the outer segments and moving away due to the spin of the galaxy as a whole). The stars on the diametrically opposite side will be traveling toward us at a higher velocity, again due to the spin of the galaxy. They will show us a blue-shift.

As the stars are revolving about the galactic center, there is a segment of them that is moving across the receding path of the galaxy relative to us. They show the true fundamental spectral band characteristic of main body of stars that are observable and are slightly red-shifted, due to the recession of that galaxy as a whole with respect to our position in space. The stars at the rotating fringe demonstrate the characteristic blue- and red-shifts relative to the mean red-shift of the galaxy as a whole as it moves away from us.

The photons emitted by these stars don't have any concept that they are moving either toward us or away from us. They are emitted in every direction exactly the same way and any color-shift is a manifestation of the observer's relative velocity and distance relationship with respect to that of the emitting source only. Did you also realize that they all also did not change in mass or physical size, nor did they flatten out due to some imaginary length contraction? Very good! That's more than many scientists have been able to understand to date.

As the galaxy moves away, it is mildly red-shifted. As it spirals with the outer fringe moving toward us, this modest red-shift is moderated by the blue-shift that accompanies the outer edge of the galaxy where those stars are moving toward us as compared with the outer stars in that galaxy.

The color is not a characteristic of the photons themselves; they have no color. The color perception is simply a mass-velocity function. In many areas of science, it may be desirable to equate the photon flux with the thermal characteristics of the emitting source. Nothing in these discussions precludes the idea that thermal characterization of the emissive source is entirely proper to do. One recalls the work of Max Planck and Sir Isaac Newton with fond regard.

The idea that photons vary in their energy content has long been established. Identifying them also, along with their thermal equivalences, is proper in helping in understanding the parameters associated with their particular sources. It is necessary for the scientific community to move away from categorizing types of radiation solely because of the manner in which the radiation is thought to be generated. A photon is a photon; a gamma ray is equivalent to a high-energy X-ray, not a photon, et cetera. A radio wave emanating from a star cluster doesn't mean there's a radio transmitter there in our normal sense of the word. The type of radiations, either electromagnetic or particulate or whatever, must be categorized by type instead of origin.

Knowledge of the source characteristics may prove to be interesting and lead to fundamental knowledge about methods for generating such radiation, but that is an aside from our discussions herein. That radio wave from that star cluster is categorically an electromagnetic wave.

As indicated in the last chapter, one of the effects of the increase in velocity for a broad-spectrum source is that the apparent energy of the detected signal will also be increasing as far as a wide-band sensor is concerned. One might think of the spectrum as photons racing past a window. As they come into this visual window, they are no longer seen as white by the casual observer but are moving ever toward the ultraviolet until they are seen no longer by the human eye.

As we know from numerous experiments in the past, photons making up the shorter wavelengths of photopic light, such as violet, have a higher energy level than corresponding photons of red. This is simply due to the increase in kinetic energy of the photon stream. If the photons have a particulate mass and their velocity increases, then due to that higher velocity we find an overall increase in the energy in the detected photon stream; this seems apparent even for the casual pedestrian.

For a light source emitting a nominal white spectrum, we can understand that for that source and that observer moving away from each other $(c - v_{rel})$, the color expected would be shifted toward the red, a lower-energy spectral band. As the source comes toward the observer, meaning of course that the light would be emitted at $(c + v_{rel})$, the nominal speed of light plus the velocity (v_{rel}) of the source relative to the observer's frame of reference, the shift would be toward the violet (blue).

Spectral shifts seen in galaxies have long been noted and identified as an indication that that particular galaxy was moving away from us. The only surprise is that we have not brought this information to bear on the idea of special relativity and the accepted concept of the speed of light as defined by A. E.

The flight path of photons should also be considered. Relative velocity, and not an electrical charge, contributes to the path taken by the photons. Electromagnetic charge is not a contributing factor in their path-of-flight deviations. If the photons are not capable of, or at least have not been shown to harbor, an electrical charge, they can pass through electrically charged atomic structures without deviation or interaction with the nucleus of ions or atoms other than by mass interactions (similar to gravitational pull) and collisions.

We have probably all seen the demonstrations of a balloon being kept afloat in a column of air moving upward against the balloon. When the airflow is stopped, the balloon falls due to gravitational forces. The air molecules are moving against the balloon with the same effective force as that of the gravitational pull on the balloon by the earth's gravity.

Now, we have seen that a small glass bead can likewise be held to defy the pull of gravity by a beam of light (photons). In the case cited, the photon beam was produced by a laser source oriented in opposition to the gravitational field of the earth.

This moves to prove the forces that are brought to bear by fast-moving particles having the ability to counter the gravity on the bead. Experiments have also been done on small droplets of oil with similar results. No electrical charge present, but solely a stream of particulates en masse, can certainly cause such an interaction. As we have already mentioned, recent experiments have also proven that the photons easily curve around large massive objects like planets and huge moons by tracking distant objects moving in back of such bodies.

There appears to be a jump of the object as it passes to the rear of a massive body; a laser sounding device would bounce a laser off of the moving body and await its return. As the object moved past the rear of this celestial body, the laser beam would show a jump in the apparent position of the moving object.

Such an observation would have been easily predicted if one were to look at the path of the laser as it passed closely by the massive body obscuring the path of motion of the object. At the most extreme point of the laser beam "bending", it is easily seen that there is a possibility of getting a return echo from the egression of the object. An apparent jump is seen and thus easily explained.

Back to our journey: our bright green beacons are in the center of the photopic range and therefore give us a lot of latitude in observing spectral shifts due to velocity using only our eyes.

Notice that the source is emitting photons in the area of a 550 nm spectrum (by definition), but as we accelerate faster and faster toward them, we see them as blue and then violet for instance, and soon we are no longer be able to visually detect them.

One might look to this also as the energy level of the light stream such as might be given by the common Newtonian definition for kinetic energy (e): proportionality to the mass times the square of the velocity, or $e = \frac{1}{2} mv^2$. If the energy level of the detected oncoming stream of light increases or decreases, one might agree that the light stream is actually a stream of particles that have mass and perform in accordance with that kinetic energy formula so well known. One might, however, wish to modify the original formula with respect to relativistic velocities such that

$$e = \frac{1}{2} m(c \pm v_{rel})^2$$

where v_{rel} is the source-to-detector relative velocity.

The idea here is to maintain that long-accepted notion that mass in motion has an energy level component. The energy is neither the same as nor interchangeable with the mass, as some people mistakenly would purport it to be, but more properly is seen as a property of that mass in motion.

We're all familiar with the equation $e=mc^2$, but perhaps it should actually be that e is not equal to but is proportional to $m(c\pm v_{rel})^2$ k, where k is a constant of proportionality and v_{rel} is the more precise value of the relative velocity of the particular photon stream between the source and the receptor rather than the familiar c, the nominal speed of light in the traditional sense. Actually, we begin to see that Newton's and Einstein's math, when changed thusly, are very similar indeed!

Let's look at the broad electromagnetic spectrum from very low frequency (VLF) radio waves through the most energetic gamma rays we have detected to date (the author certainly allows for the existence of a much broader spectral/energy range for the electromagnetic family).

We find the human photopic region nestled somewhere in a central position. Now, there is reason to believe that the photon region is of a pedigree entirely different from the electromagnetic family in its construction and interactions with matter. It is also highly likely that other particulate-based radiation may be likewise misidentified as being electromagnetic instead of particulate.

Now, many of the characteristics of photon and other relativistic particulate radiation behave in much the same manner as electromagnetic waves. Much has been done in the past to try to ensure that all these behave in the same manner. A.E. even took pains to modify his fundamental mathematical analysis on occasion to accommodate the subtle differences not otherwise manageable.

Photons don't travel in waves! Photons, along with other similar phenomena, are actually particulate streams with energy levels that we have tried to define as having a wavelength. The wavelike properties can be explained by particulate interactions, in part, and wavelike when in sufficient quantities to combine and form the wave action described.

Much work, of course, will need to follow to fully understand the extent of resultant reactions and contributions to particular fields of physics that this new revelation leads us to.

For over a century, men have argued about light being particles or waves. In many experiments, light would first seem to act like one, and then in another experiment, like the other.

Dude, these waves are awesome!

PHOTONS ARE PARTICLES

Perhaps if we could simply look at these photons as the extreme of smallness of size and therefore of an extremely small mass, we could then understand that the trillions and quadrillions of photons moving together could easily be seen to act as the front of a wave.

While water molecules, water, as well as waves of it, are not exactly analogous to our efforts here, we can look at the huge volume of water molecules in a moving stream of water; the water moves around objects and then the molecules come back together. Small particles such as photons could likewise easily move around objects and, through attractive and outside forces, be pulled back together in a flowing stream like water.

They would certainly appear to act as a stream and like waves in many situations. Photons are minute particles that have mass and, therefore, by virtue of their velocity, have energy. The energy is proportional to the mass times the velocity squared, along with a variable that identifies the relative velocities between the emission source and the observer. Some adopt a temporary and loosely embraced analogy of the photons' wavelike behavior as being like the waves formed of water molecules!

Now, we know water is certainly made up of small molecules, but they coalesce to then be considered as a liquid that behaves with wavelike actions. Our photons likewise coalesce to be seen as a stream of light, and light behaves in large part with wavelike actions.

If we now understand that the wavelength or chroma character of this light stream is a result of its energy level, which is a function of its velocity, relative to us as observers, we'll also see that this idea of energy level between the moving particle and the receptor (detector) can be carried throughout the entire visible light spectrum and beyond. Again, we are reminded of the black-body radiation models that are so well known.

Our ability to detect such particles is limited, in great part, by our presumptions that they are waves. We need to expand our sensor/detector capabilities and begin anew to ascertain the full detection range of the particulate/electromagnetic energy spectrum.

We're not detecting colors; we're detecting relative energy levels. The minute size of such energetic particles would react on an atomic level with any elements and on a molecular level with compounds, crystalline, and complex matrices. The idea of photons having a wavelength or frequency of oscillation has no validity in light of this (yes, a pun). The fundamental fact is that we detect the photons as an energy level of moving particles.

Consider now an interesting aside: if the human eye detects photons but an observer is color-blind, does it mean that he simply can't discern the energy differential between the various photons that are encountered? This might put an entirely different spin on the research that might go into resolving the mechanism of, and eventually curing, color blindness!

Simply having an extremely high energy level particulate doesn't mean that a particle of mass has a frequency of oscillation or a wavelength component.

Photons are not a subgroup of the electromagnetic spectrum of radiation, although there are similar actions and reactions as such. It is necessary now to reinvestigate the various radiation types carefully to determine the true nature of those particular radiations.

Let's try a bit more digging to see what we can come up with. What mechanism do you think could cause these color differences in the photon streams? Well, let's take a closer look at the photons from a quantum mechanics point of view.

Many decades ago, Max Planck worked on the various colors of light being emitted from black bodies; that is, they were emitted from the body at different temperatures and different energy levels. This is not quite the same as our observation methods, but they are related. It turned out that the output spectra of photons could be correlated to the temperature of the emitting body.

Many years ago, it was also found that if metal substrates (and other items) were bombarded by photons, they would emit electrons as a result. These experiments found that photons knocked the electrons off the metal surface. It turned out that electrons had been knocked off of the metal surface when struck by photons of a blue to ultraviolet wavelength. These photons had a higher energy than red ones, and the higher energy levels manifested themselves as electrons being knocked off by these photons.

If you remember, Isaac Newton came up with lots of ideas about matter, energy, velocities, mass, gravity, and the like. One of his equations was that of particle energy as a function of the mass of the particle and its velocity.

A number of scientists over the years have worked to challenge and prove or disprove the validity of Newton's work in this area; it seemed not to fit precisely with the newer work done in quantum mechanics. Einstein came up with a derivation that simply removed the ½ status of the mass portion of the formula and replaced the velocity with the velocity of light (again, when he thought it was to be a universal constant and defined a cosmic speed limit).

This sounds a lot like the photons we've been observing. If these small photons are tiny bits of matter, then we can understand that the blue ones are moving faster and are therefore more energetic. We know from the laboratory experiments with the metal and the electrons being knocked off and from our own observations that the blue photons are moving faster than their red brothers and are certainly more energetic because of it.

The differences in photon velocities may be of a minor difference to our way of measuring them, but that small difference is indeed valid and of significance to other physical phenomena we have studied and have yet to study.

7

PAST ERRORS REVISITED

Experimental convolutions, diversions in the train of thought, and non provable postulations, along with forceful, charismatic, and convincing presentation, all converged to produce impressive erroneous results.

The idea that dimensional and time distortions occur, or further, aided in proving null results of an experiment was both unnecessary and fraudulent. It has led us down a path that has taken decades from which to recover. Many scientists still stride down paths that include the space-time continuum, string theory, space foam, black holes, and worm holes, as well as multi-universe concepts (do you suppose they should be called "multiverses" instead of universes?) and time travel.

Let's take a look at what we have been told up until now. We've been told that if we had a spacecraft of our very own and went on an excursion for an afternoon (or perhaps for a few short years) and zipped around space at the speed of light (c) or a bit less, we would experience a physical foreshortening on our spacecraft and a time dilation, according to arguments by Lorentz and Einstein.

If I had a twin and he had a spacecraft exactly the same as mine, and we went out in opposite directions or perhaps only he did, then we would end up aging at different rates! A potential cosmic "fountain of youth"?

REALLY?

In view of much of the speculation of the past, we have also assigned a fixed velocity to photons and then used that speed to calculate distances in the cosmos. Perhaps what we have done is to generate calculation errors as large as space itself.

What would we say if we found out that the actual speed of light was not a constant, and a small error in using it as such would cause light-years of variance in astronomical calculations? It occurs to me that the size of the universe might not what we've been led to believe. This also means that the universe is not the age we've been led to believe.

Imagine the multitude of ramifications these small discrepancies can possibly cause. It's time to stop and revisit all of the fundamental research that has been based on erroneous equations, assumptions, and suppositions. If we now concentrate on changing our outlook and measuring approach and begin by detecting photon energy levels, we may be able to stride forward with more speed and positive determination than ever before.

More care must be given to comparisons between various types of radiation, wave theory, sources, and detectors used throughout the sciences. Do the source of the radiation and the method of its generation define its type, or is it the physical interactions that we are able to discern?

Through various studies, we've found that photons can and do cause electron emission from metallic substrates. It has been found that the blue photons are higher in energy level than the red ones. We attempt to define work functions of elements and materials to predict future reactions to photon (and other particulate) bombardment.

If we look back at Planck's brilliant work with an eye toward energy levels instead of spectral characteristics, we would be able to more clearly understand that the spectral assignments we impose on photons is simply derived from the use of our biologic receptors (eyes) and in reality have nothing to do with frequency of oscillation of the photons themselves.

Perhaps it's time to revisit all of the mathematical calculations that are based on the wavelength of light and replace them with the same calculations, using photon energy levels instead. We can use the fundamental energy levels derived from solid empirical data gathered over the decades and thus affect a positive change that will form a solid base upon which further research can flourish.

Let's ask another potentially embarrassing question: if one uses the speed of light c as a fixed and absolute value and then uses that to gauge a distance (such as from the earth to its moon), does that make the distance known?

If we measure using such soft numbers in our work, we may be in for harsh and eye-opening contradictions in our results over time. Is the universe 13 billion years old? Is it as large in size as we have calculated it to be? What precisely is the mass of a photon? How can we correlate the photometers we've been using for years with the energy data in order to obtain a more solid base for our future photonics work?

There is no reason why any of these questions should go unanswered, but they should be answered with certainty and a high probability of precision. This effort must be undertaken, and the field of science utilizing such data can then feel confident in the product of its work.

What we have just done here is to take the work of Albert Einstein and his conclusions and shown them to be incorrect. It's time for the physics community to come to terms with the overwhelming evidence of decades of experimental results, modern instrumentation, fresh ways to looks at the vast pools of evidence, and find that the "truth era" of physics has begun. Many of the problems with the theories and their steadfast followers have been that many followers were simply that; followers.

Many would go to great lengths to prove the tested theories to the extent of numerous derivations of mathematical formulae and convoluted arguments to make them work out. Lots of folks simply couldn't understand or take the time to thoroughly understand the theories and simply accepted what was taught in the universities; after all, it was the university, the center of truth and knowledge, right? Who would dare challenge Professor Einstein? Well, actually, lots of folks have challenged the work and its conclusions over the years.

Many have been laughed at and attacked by his loyal followers, many have not been able to clearly and persuasively communicate their ideas, and even more have questioned in privacy, either not wanting to rock the boat or just a little unsure of the validity of their ideas as compared to Einstein's. What a shame.

PROFESSOR KNOW IT ALL

The conspiracies against these changes are often unworthy of professionals. There have even been publishers who refused to print or publish anything in opposition to Einstein's work, some of them for the same reasons cited above, and some because they just wanted to hide their heads and not lose face because of their following along for so many years. Preposterous! Not letting the truth out because of simple vanity! Their inaction certainly is one cause for keeping the false postulations alive for such a long time.

What other conclusion might we come up with as a result of our simple experimentation? We understand that the photons being emitted from a source such as our sun are emitted with no concept of color. When the photons leave in every direction, they all look the same. The ideas of color or spectrum or frequency or vibrating photons or wavelength have no validity for these simple particles.

What we are observing is simply the energy level of particles in motion that possess a minute mass. As the energy of a particular photon is higher, we see it as bluer, and it is therefore moving faster than its red brethren. But wait! Some experiments in the past have tried to understand that all photons move at the same velocity, and as proof of this they begin to discuss a lack of phase shifts in their test results.

ALL PHOTONS ARE NOT CREATED EQUAL

They maintain that if the photons were moving at a range of velocities from some known source, the arrival of these photons should be seen as a difference in overall transit time as a phase shift.

Perhaps the shift is not enough to be easily perceived; perhaps it doesn't exist. This is a question for the future experimenters to answer.

8

RELATIVITY ACTUALLY

If we accepted Einstein's idea that the speed of light is a universal constant and is fixed, without regard for the velocity and direction of the emitting source, with respect to the observer's position, then as our fellow traveler precedes us and we glide along, say at 1.0 c, he would lose sight of the photons emitted by our spacecraft.

Still, with our ships, as we motor along, if the velocity of the photons is dependent on the velocity of the emitting source, we will detect the light from our companion spacecraft as being green since we are moving toward the emitted photons at the normal relative speed of the emissive source.

By we now understand the idea that the velocity differential between the emitting source and the observer's frame of reference does indeed have a significant effect on the observations; the pilot of the other spaceship will also see us emitting a green light.

If he accelerates to a speed faster than us, and away from us, we will see him as red-shifted. He will see us as red-shifted as well. If, instead, we accelerate toward him, he will see us, and us him, as blue-shifted nominal photon behavior with respect to our frames of reference.

Both the other spacecraft and ours can be judged with respect to the same frame of reference as we travel through the local space partition.

If we now move up beyond the speed of light with respect to our LSP frame of reference, our new friend in the other craft, in looking out his back window, will no longer see anything close to his path of travel other than our craft. He'll be able to see us since we are emitting green light in the same frame of reference in which he resides. As we both continue to go faster a "cone of blackness" is enveloping that which has been left behind us.

If our friend's little craft now moves up beyond the speed of light with respect to our frame of reference, we will no longer be able to see him since the photons he emits will now be moving away from us.

Such is not the case for the rows of beacons showing us the way, however. What happens if we watch an oncoming beacon? It appears as shifted toward the blue, and as we pass by, we see the light emission move through the spectrum to show the red shift as it goes toward the rear of the craft and then into the black cone toward the rear. It may be very important for us to install sensors aboard that will look well ahead of the craft and span a wide energy range well beyond the capabilities of the human eye.

The next day as we travel onward, we encounter another identical craft coming toward us on our same path of travel (but far enough off our path to allow for no collision). We're speeding along at a 0.8 c and the other craft is just now smoothly accelerating up to 0.8 c as well.

The small green lights we have on the front of our ships are working properly. The questions must be asked: "How do we know we're traveling at 0.8 c?" "How does he know he's accelerating to 0.8 c?" With respect to what? Maybe there are more buoys present in this local space partition and we can judge our speed relative to them.

It turns out that we always assign our speed with respect to the static LSP as a normal routine. It is common practice to judge movements and velocities with respect to our own spatial frames of reference, but now it becomes clear that all velocities and movements must be adjudged according to relative space. A clear and specific definition of that space is required in order to advance our space travel concepts beyond our solar system and galaxy and into deep space.

Moving on again, as we pass each other we will not, however, see the light coming from his rear light and he will not see ours again because we are moving apart at too high a velocity. But *wait a minute*! Haven't we just experienced a spacecraft moving past us at 1.6 c (or us past him, with respect to our frame of reference)? Right!

Can we draw from this that light, particles, and bodies of significant mass can move well beyond the speed of light relative to our frame of reference? Of course we can!

All of this, of course, requires that the speed of light be relative to the emitting source and not a fixed universal constant (and limit) in the universe as some would have us believe. Now, if the other craft were stationary in the local reference frame, we would detect his light as blue-shifted (beyond visible to the human eye) due to our velocity approaching him (still at 0.8 c). He would likewise detect our light as blue-shifted.

If we stop and as we see him accelerating toward us with respect to our new frame of reference (that is, of the local space partition), his light would appear to us to move toward the violet and then disappear into the ultraviolet and far above. This would also be the case for many incident photons reflected by his craft in our direction, so essentially he is now invisible to our eyes. All photons emitted by him prior to his achieving c_{uv} speed will be seen by us. Defined here, c_{uv} is the velocity of photons as they move into the ultraviolet portion of the spectrum. If we have a detector that is capable of detecting all wavelengths of particulate radiation, we could then still detect his presence. He is likewise able to see us, or not, for precisely the same set of reasons.

It may be a bitter blow to some, but our frame of reference is not *the* sacred preferred frame that the universe is constrained or in any way controlled by. So much for the long-held idea that the speed of light was fixed, limited, and, indeed, defined by a limit of relative velocity for all other bodies in motion; that's clearly not so.

While many of our ideas are outside the ability of terrestrial laboratories or even our current space-borne equipment to validate or disprove (at the present time), they are compelling and certainly food for thought. We are, alas, at this point in time, limited to thought experiments, in great part, on the subject.

Observations made over the years and made known to the scientific community, and more recently the Internet community, bring us all together in a loosely bound coalition that moves ever closer to the understanding that we all so deeply desire. We are required to explore all the possibilities and probabilities and listen, without bias and with respect, to others who may have a helpful idea or a different perspective from our own.

We question the presently established laws and rules that have been formulated with heretofore limited resources and information. As scientific understanding and technology move forward, we will uncover more and more discrepancies in the old ways and learn to embrace the new.

Now, as we go back to our cozy spacecraft, we worry, and with good reason. If we are traveling above the speed of light with respect to the local space partition, or someone else is and is moving in our direction, how we can be safe from collision? What a collision that would be! If it was a manned craft (and we were moving much slower than the speed of light with respect to the other craft), then we both could take evasive action, but particles and non-piloted massive bodies wouldn't do likewise.

Now would be a good time for someone to invent a device for warning of, or at least being able to detect, objects moving faster than the nominal speed of light. (But once again, with respect to what frame of reference)?

Any active system like radar would have to have a device to send and receive information that is greater than twice the speed of any object that may be closing upon us. Perhaps a device will be developed that squawks a warning signal from a fast-moving manned craft toward its path of flight. Such a signal again would have to travel faster than the craft and be detectable as well to allow time for reaction and maneuvering for avoidance. Again, that's a tall order. But we are getting way ahead of ourselves.

We just learned to go that fast. As we have seen over the past several decades, there's not too much traffic in our part of the universe, and so far we don't know if "faster than light" bodies have ever been detected out here; if they have, no one's talking.

One would certainly have to utilize sensors that detect well beyond the deep-ultraviolet portion of the spectrum. Perhaps we should recommend that we begin such observations if they are not currently underway.

We understand that the photon we detect has a velocity at the moment of emission and is also dependent on the velocity and direction of movement of the emitting source with respect to our frame of reference. We should consider what velocity a moving particle or massive body would need in order for the spectral shift to exceed the photopic range of our eyes' perception. We move into the well-traveled domain of Doppler shift for an explanation of what we may see and thereby try to define our limitations.

Let's suppose the photopic range for light is from 400 nm (violet) to 750 nm (red).

Let's also suppose that the source of the light is now moving toward us at a rapid velocity, say 0.31 c. Let's propose that the source of the light would emit only in the center of the photopic region and appear as the green light if not moving with respect to our frame of reference. We'll use a wavelength of 550 nm for our calculations. What is the apparent spectral shift we could expect? We'll use the number of 300,000,000 m/s for our nominal speed of light c.

Now, using the classic Doppler shift equations:

$$v_{in} = v_{out} \sqrt{\frac{1 - w/c}{1 + w/c}}$$

where v_{in} is the wavelength as seen, v_{out} is the wavelength emitted by the source, and w is the velocity differential between the source and the observer.

We expect a shift toward the violet end of the spectrum to a wavelength of approximately 400 nm, or just at the limit of normal human vision. Likewise, if the source is moving away at 0.237 c, we see a shift of about 150 nm toward the red to approximately 700 nm, the upper end of the human vision range.

Let's go back to the two spacecraft moving toward each other, but this time they are much larger craft than before. We now find we have enough room inside to set up an experimental apparatus!

There is a design for a reasonably small speed of light measuring apparatus that will fit comfortably inside, and we can conduct a few experiments.

Our first experiment is to just check out the apparatus to ensure that it's functioning properly, so we fire it up and get the same results we had gotten when we first tested it back on Earth. This conforms to the idea that in all frames of reference all laws of classic mechanics hold, even as we speed through space taking along our own frame of reference with us.

Now, we find out that the other guy has just done the same experiment, and he also has gotten the same results! Yes, he has his own frame of reference as well, and, yes it is supposed to work that way, but if we are moving toward each other at a high velocity, doesn't that mean that the speed of light as measured in both craft are moving at c with respect to their individual sources and test apparatus?

Since that's true, then the speed of light inside each craft is certainly not moving at c with respect to the other craft. It's also not true for some imaginary frame of reference where some interested observer might be standing and watching the goings-on. The firm conclusion is that the speed of light, c, is not a constant in the universe but simply a relative benchmark for whatever frame of reference it's being generated in and is independent of the frame of reference in which it's being observed.

The observer will perceive the spectral character of the light being emitted, depending on the relative velocity between the emitter and his sensor, differently, without doubt, from the fellow in the frame of reference in which the light is being emitted.

This will also hold with the examples of chroma shift of celestial bodies as the velocity between them and the earth varies. The chroma shift is not indicative that the wavelength is changed but is simply a difference in the velocity and therefore the energy of the photons we're observing, which varies with respect to our frame of reference.

A simple observation at this point would help a lot! We can detect and understand the photopic spectral shift characteristics of celestial bodies at great distances from our location, and they are indicative of those bodies moving with respect to us at a velocity significant in comparison to the nominal speed of light, c.

We can also clearly understand that this means that the velocity of the light is not a constant with respect to a *preferred* frame of reference outside of the emission source. Indeed, the speed of light noted is a measure of its velocity and energy as a function of the relative positions, velocities, and directions of relative travel between the emission source and the observer!

Do I wonder what that does to the Special Theory of Relativity? No! I don't wonder about that at all; SRT has never been valid. The idea of the poor clocks that couldn't keep precise time, the twins that aged differently, the rods that shrunk in length, and the particles that grew more and more massive with speed are all components of preposterous theory that is proven to be a longstanding illusion and has been widely taken, as if in an enchanter's spell, by the physics community in general.

There have been learned men throughout the years who have argued against these theories, but their numbers were insufficient and their combined voices not nearly loud enough to stand against the overwhelming din produced by the parrots and other less enlightened gentlemen that all fell into lockstep with the master, A. Einstein.

It is now clear that all observers have their own frames of reference and if all are located together on a single platform (read: Earth), they will then have the same common frame of reference. If they are moving about either by locomotion, aircraft, boat, etc., they then again have their own unique frames of reference. Every movement that one observes or that is observable by someone is therefore, *by definition*, uniquely relative to that observer. Each observer can correlate his data with others of the same or separate and distinct frame of reference mathematically by a transform series, similar to (but not) the Lorentz Transformation Equation set.

Several physicists have strongly disputed the Lorentz Transformation Theory and have indeed proven that it does not hold; there are mistakes in the concept, the logic, the mathematics, and therefore the conclusion. What does this do to cosmological calculations as to the size of the universe, its age, and SRT? Let me refer you to the work by Dr. Hans Zweig once again and strongly suggest you read his work, *Relativity Unraveled, A Question of Time* (www.aquestionoftime.com). Dr. Zweig concludes that the age of the universe is about half that commonly accepted by astrophysicists.

If one ventures out using Euclidean space and Newtonian theory, then the resultant age comes out to about 6 to 7 billion years or so. The popularly held age is believed to be between 12 and 13 billion.

9

THE TRAIN, THE TRAIN

A friend and colleague of mine, Dr. Hans Zweig, has brought forth some interesting and thought-provoking ideas concerning the Special Theory of Relativity (SRT) as published by A. Einstein some 100 years ago. Dr. Zweig, with insight and by use of his ample persuasive logic and skill as a mathematician, has shaken the world of SRT and indeed has succeeded in casting great doubt on the SRT and Lorentz Transformations (LT). The resultant theories form a basis for reconsideration in the realms of cosmology and relativistic physics. No doubt he'll go down in history as one of the great minds that sought to enlighten the community through long effort and critical review.

In his book,[1] Doctor Zweig put forth a thought experiment of a train using a very long track (embankment, keeping a similarity with A. Einstein's work in order to draw a comparison), a train that was 300,000 km long and traveled at 150,000 km/sec. On this train was a mirror attached to the front with which reflected beams of light could be sent rearward to a detector located on the rear of the train.

One beam of light would be sent from a pulse source, A, located in the rear of the train adjacent to the detector, and another pulse source, B, is located on the embankment beside an observer. The idea was to discover what happened to two pulses of light (both blue) that were simultaneously initiated when the rear of the rapidly moving train (along with the detector) came adjacent to the observer's position on the embankment.

MIRROR, MIRROR ON THE TRAIN

Dr. Zweig offered several possible answers to the experimental results depending on the approach of the scientist conducting the experiment, and then included another alternative view brought by a colleague using the length dilation of Lorentz and SRT. In testing the experimental results, we find:

In scenario one, pulse A: It quickly became clear that the pulse from, and on, the rear of the train would reach the mirror affixed to the front of the train and reflect back to the detector at the rear after a total of two seconds of elapsed time. The time of light transit totaled two seconds, and in this length of time the train moved a total of 150,000k/s * 2 s, or 300,000 km.

In scenario two, pulse B: The light pulse is emitted from the source on the embankment and is sent down the track in pursuit of the mirror mounted on the front of the train at 300,000 km/sec with respect to the embankment frame of reference (but only 150,000 km/sec with respect to the train and mirror frame of reference). This pulse reaches the mirror after two seconds of elapsed time, where it is then reflected rearward toward the detector located at, and on, the rear of the train.

Now comes the potential for argument and different ways of looking at the reflected light pulse, B. In order to clarify the true effect of the mirror as it moves along the track, it helps to "see" what the light pulse source looks like to the driver (and therefore the sensor at the rear of the train), who is on the train and looking at the pulse source via the mirror.

Argument one: For the mirror mounted on the front of the train and the light pulse emitted from the embankment source, the source is seen as receding at the same rate of speed of the train, or 150,000 km/sec. The light pulse coming toward the driver is a red-shifted pulse (meaning from its originally emitted blue color, it is shifted toward the red end of the photopic spectrum) due to its effectively reduced speed. The result is that the driver will see the blue light pulse as red-shifted by either looking into the mirror to see it or simply upon turning and looking back as the pulse arrives.

Upon reaching the mirror, the light pulse is now reflected at the same speed at which it overtook the mirror and moves rearward toward the detector at a rate of 150,000 km/sec with respect to the train, driver, and detector.

The red-shifted blue light will not shift back to the originally emitted blue. This makes the return trip time to the detector two seconds, and the distance traveled would be 150,000 x 2, or 300,000 km. The total train movement would be 600,000 km!

Argument two: Remember, the light pulse is being generated on the embankment for this argument. The reflected light pulse essentially uses the mirror as its source and, in moving rearward, travels in the train's frame of reference at 150,000 km/sec.

With the rear of the train fast approaching at 150,000 km/sec, this results in a total velocity of convergence of 300,000 km/sec and a total travel time of three seconds for a total distance traveled of 450,000 km.

Let's have a look at argument two above. The mirror magically becomes the source for the light pulse, and the light is emitted at 300,000 km/sec in the reference frame of the train. The detector at the rear of the train will see the pulse exactly one second after the reflection, and therefore the total time will be three seconds. This would also mean that the light seen at the detector would then again be the original blue pulse.

Argument three: For a moment, let's put the reflector on the embankment adjacent to the point where the mirror on the train will be at the same instant that the pulse of light arrives. The mirror is now in the reference frame of the embankment, situated at the exact point where the B light pulse and front of the train reach at the same moment in time. The mirror reflects the light pulse back toward the detector that is on the rear of the train.

This changes the mirror's function with respect to the light pulse in that it is no longer traveling away but sits stationary in the reference frame of the embankment awaiting the light to reflect.

The mirror must then do as stationary mirrors do and simply change the sign (the direction) of travel of the light beam. Now we have the light pulse moving rearward at 300,000 km/sec and the train moving forward at 150,000 km/sec for a total transit time of two and two-thirds seconds. This yields a total travel distance of 400,000 km. If the mirror is on the embankment instead of on the train, and the driver of the train is looking into it to see the source of the pulse of light coming, the mirror would itself appear to be moving toward the driver, as would be the pulse of light reflected by it as the front of the train passed it.

It is perhaps easier to visualize if one considers the mirror to be as large as a roadside advertising sign of huge proportions alongside the tracks. The image seen in the mirror would not be receding but, instead, advancing toward us. We would be effectively moving toward the source of the light pulse. The reflected light would be blue-shifted, or from its original blue (into the ultraviolet), as seen by an observer (the train driver) in the reference frame of the train.

Upon reaching the detector and as seen by a casual observer on the rear of the train, the pulse is then ultraviolet; but as seen by an observer on the embankment the arrival of the light pulse reflected by the mirror and arriving at the rear of the train (detector) it would still be a blue pulse.

The time of travel for the rearward portion of the journey would be 0.66 seconds (300000 km/sec velocity of the oncoming pulse and 150,000 km/sec for the forward velocity of the train), and the travel distance for the train would be 100,000 km. This, added to the two seconds of time and 300,000 km traveled during the "overtaking" phase of the experiment, results in a total of 2.66 seconds of time and 400,000 km of travel of the train.

It is easily seen that the location and "status" of the mirror and its motion in the exercise is critical as the location of the observer. It also requires a single conclusion for the situation where the mirror is mounted on the front of the train (argument B-1), that four seconds is the total elapsed time for the light pulse from the source on the embankment to reach the detector on the rear of the train and 600,000 km in train distance traveled. No other answer is possible.

It would be somewhat desirable for one choosing a wavelength for the light pulse being transmitted to choose one in the middle of the photopic region (perhaps green at 550 nm) so as to allow for shifts in both directions to be more easily observed during the course of the experiment.

Now, those who want to include either time dilation or length contraction should consider this: if the embankment would be situated in the vastness of space and the train in casual contact with the rails, as would normally be the case, an observer on the train would see the embankment moving rapidly past while an observer on the embankment would see the train moving rapidly past.

Which of these frames of reference do you suppose is the foreshortened one? Can you actually believe that there is a preferred one? If you are a scientist and answer in the affirmative, I suggest you review your physics texts or perhaps change your profession.

Perhaps another example will help. Two identical spacecraft are moving through space toward each other. Both see the other as different from their own craft due to the Lorentz Transforms. Which of these craft actually remains unchanged and which does not? Clearly, if we buy into the Einstein and Lorentz arguments, we pick ours as the preferred frame of reference. *Wrong!* Each craft has its own frame of reference, and we can even define a local space partition (LSP) with yet another frame of reference.

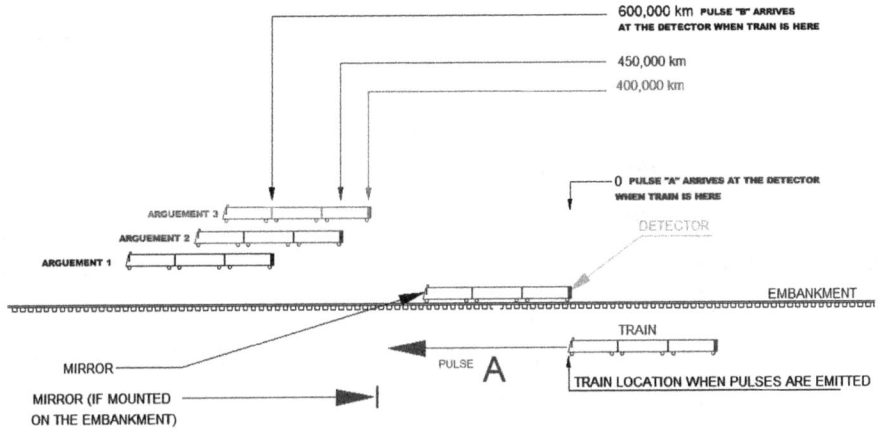

1. TRAIN IS 300,000 km LONG
2. TRAIN SPEED IS 150,000 km/SEC.
3. TRAIN POSITIONS ARE SHOWN WHEN
   REFLECTED PULSES (B) ARE DETECTED

600,000 km PULSE "B" ARRIVES
AT THE DETECTOR WHEN TRAIN IS HERE

450,000 km
400,000 km

0 PULSE "A" ARRIVES AT THE DETECTOR
WHEN TRAIN IS HERE

DETECTOR

ARGUEMENT 3
ARGUEMENT 2
ARGUEMENT 1

EMBANKMENT

TRAIN

MIRROR
PULSE A
TRAIN LOCATION WHEN PULSES ARE EMITTED

MIRROR (IF MOUNTED
ON THE EMBANKMENT)

TRAIN GRAPHIC ILLUSTRATION

Any amount of mathematical manipulation to the observed data done in any of the frames of reference in order to correlate data with another one obviously does not change the observed occurrence but simply seeks to define the perspective of the observing individual.

For too long a time, many timid scientists have looked past the illogical and questionable application of mathematics, incorrectly applied, and never blinked. It's time to come up to the table and have a good look at the scenarios put forth herein and come to the conclusion that in each frame of reference, the rules of physics hold as in any other frame of reference, and there is no preferred frame that will make the others faster, slower, longer, or shorter.

There are perspectives from one frame to another that require that an observer in one frame modify his observed results because of relative motion, angulations, or variation in distances during an observed event, *but that certainly doesn't change the reality within the frame of reference of the event being observed.*

Think of another idea this way: if we observe a cluster of stars as red in color, are they red with respect to their frame of reference or just ours? Maybe they're blue, red, green, or ...?

The coloration we observe is relative and based on other parameters as well, not spectrum, and they're certainly not red because *we* see them as red. Our observation has nothing to do with their state of being but only with our perception of their state of being.

Now we can understand and move to agree with our colleagues like Dr. Zweig and others who so carefully have studied, challenged, and argued against the validity of the LT and SRT over these past years, just to be rebuffed by those who would not see for fear of making the wrong decision against the popular and famous A. Einstein.

1. *Relativity Unraveled, A Question of Time* by Hans J. Zweig, ISBN 3-9807378-4-5, 2004

- I 0 -

WHAT CHANGES?

With the preceding information, we can now look back at some of the work that has been used to support or even been based on the original Doppler application and those based on the "fixed" speed of light limitation held for so long a time.

It is clear that Einstein's Special Theory of Relativity was erroneously based on the supposition of the constancy of the speed of light in the universe. His assertion that light was held as a universal constant without regard for the relative velocity of the emitting source was clearly wrong. From our analysis of the nature of photon emission and detection, it is seen that the relative velocity between the two bodies is a key component in defining the energy level of the detected photons (which equates to what has been historically been termed a wavelength of color).

In Einstein's paper that included the subject of space and time in classical mechanics, one might recall a particular observation and subsequent conclusion that a pebble moving from point A to point B did not have an independently existing trajectory but instead was defined by the observer's point of view, which is the observer's coordinate system. This, of course, is ridiculous on the face of it.

It is simple enough to understand that as the pebble moves from point A to point B, it does so in its own frame of reference. If someone watches this pebble from some other location, the pebble still moves as before without regard to the observer. Its path may clearly be defined in its own reference frame.

Consider the pebble as a miniature world moving about in space; that world in motion is not in any way constrained or changed by someone observing it from afar. The earth's movement, for example, is unaffected by some person light-minutes away watching it move about through the solar system.

We can then describe the frame of reference we are attached to and apply a mathematical transformation to describe our observations. The path of the pebble will obviously be different as viewed from different perspectives but has a finite trajectory of its own that is independent of any coordinate system we wish to construct. Any correlation or transformation we perform on our data has no effect on the physical being of the pebble or movement thereof.

The idea of a coordinate system that magically constrains the movement or limits the motion of a body being observed is pompous indeed. Any coordinate system is artificially constructed for the advantage of the person using it in order to better describe events he may observe and wish to relate to others using that system or frame of reference.

This holds true for the concept of time measurement as well. The observer watching the movement in his contrived coordinate system uses a device to measure the time of travel as well as the coordinate length measurements to more fully define the movement observed strictly with respect to his frame of reference alone.

The velocities observed in the movement of the pebble may vary with time and displacement and the observer, who with his measuring devices, can define that path for his own use and communication with others. The displacement in the coordinate system he has constructed and the time of the movement that he has "measured" are fabrications of his observations alone. They are not necessarily identical to nor do they define the motion of objects in the reference frame of the pebble.

This is where many folks get off track. They believe that their coordinate system and clock are *the* coordinate system and clock! Nonsense! This overwhelming arrogance has provided an obstacle too long-standing! We are not the center of the universe, and our methodology and resultant parametric data must be related to others as simply relative to us, as seen by us, not as absolutes in any sense of the word.

Any effort to describe the object and its movement in its particular coordinate system can certainly be discussed with persons in or part of another system with a simple correlation within the several systems themselves. Any correlation functions will have no effect on the object or its motion. As we move along the experiment, we now find a need to discuss time.

In setting up this discussion, two identical clocks were used. This, of course, is untenable if aliens from a far point in space need to discuss that subject with us. We should use a more universal concept of time and method of measuring it. Einstein linked the Galilean and Newtonian systems together and then declared the Galilean system erroneous since it was affixed to the earth and the earth was moving about in the universe.

The system was not useful for discussions where celestial objects were involved. Thus, by association, he declared the Newtonian system of mechanics void as well.

In order to free the Galilean-Newtonian system from being earthbound, one simply declares that the coordinate system is now defined to reference distant stellar objects that are at extreme distance with respect to the area of interest for his particular experiments.

There will still be rotational and translational components with respect to cosmic bodies, and this cannot be avoided because the universe is in an ever-expanding mode and all objects are moving about within the overall expansion.

Any observed movements or activities must, therefore, be related to specific points or objects and these relationships made a part of the data set that is valid for only a brief time indeed!

Since the entire cosmos is in constant motion, it is perhaps more sensible to refer to our LSP as that region of interest for anything needed for our esoteric thought experiments. The idea that we can define anything of significance at far reaches of the cosmos is human folly indeed. We occasionally get full of ourselves with arrogance and actually think we can watch, formulate, postulate and conclude facts and laws for the far reaches of the universe.

A fool might wish now to define that one coordinate system in particular is *the* coordinate system and that all physical laws that dictate mechanics in his system are valid only for that system, awaiting only a refinement by relativity calculations. In fact, the motions and objects pay no attention to any coordinate system but move according to the universal constraints placed on them by natural forces, not by man.

If an object is emitting photons as it travels through space, the velocity of the photons with respect to the emissive object is approximately c, period. If another craft is moving about with us and is also emitting photons as it goes, they, as well as us, will also see the photons the craft is emitting to be traveling off at c.

If one of the objects (now a spaceship for convenience) wishes to describe the movement of the photons emitted by the other craft, it simply describes its observations with respect to its own coordinate system, whose origin must, of course, be the observers' craft itself.

In order to describe the photon movement to someone on the other craft or any other person on any craft or planet, one simply describes his observations and then defines his coordinate system and describes that to him as well. The relative motion between the various systems will certainly modify the perceived action of the photons as far as how that person would see the event but certainly doesn't perturb the photon stream itself. Yet another attachment of these photons has been attempted in firmly linking them with the studies done on electromagnetic radiations and optics, which some physicists believe are equivalent and must obey the same "laws." Such is certainly not the case.

Another slight of mind occurred in trying to describe the relativity of simultaneity. One must keep in mind that the events that take place do not do so in the observer's coordinate system; they occur freely in independent space. Occurrences in this space must be related to all coordinate systems for any discussions as to time, place, and movement, or to other occurrences observed. Any events will occur with their own physical constraints and properties, not those of any observers.

Should one wish to describe what he has observed and therefore what conclusion he has made must do so with respect to his particular coordinate system only. If the same laws of motion, force, time, and mass are used throughout the cosmos, then a simple construct of relativity can be made once the relationship of any associated coordinate system is defined.

A need to find that time for coordinate system A is the same for any other coordinate system (B, C, D, etc.); it is simply a matter of definition and nothing else. We have the need to discuss in like terms in order for the discussions to have any sense of validity. One second for us must be the same for the party with which we discuss the event, or a correlation must be established that will make it so. One meter, likewise, must have the same definition of dimensions, as well as time; definitions of all coordinate systems require consistency or discussions will ultimately fail.

Now we begin to look at objects and events with respect to two distinct coordinate systems, K and K'. If the object and event takes place and is described in system K, then a transform such as accomplished by Lorentz will modify the results of observation about the object and associated event as if seen by system K'.

Such a transformation, while incomplete, is useful but must be used with awareness of limitations. Firstly, it must be realized that the object and event do not exist in the observer's particular coordinate system but are only observed from it, and secondly, the transformation does not include extended motion that would include geometric distortions or changes in perspectives such as the relative angle of view seen by the observer in any coordinate system. Some will have you believe that the length of a meter rod is no longer a meter in length, that time of movement is not really the time of movement but changes because the observer sees it change. These aberrations are fundamental to Einstein's Theory of Special Relativity, and the transference of these physical and temporal characteristics to the object and event is completely preposterous.

Based on derived formulas utilizing the Lorentz equations, Einstein then determined that the speed of light, c, was a limiting velocity and a fixed quantity throughout the universe. This leads to the idea that no object in the universe can exceed the speed of light. How utterly arrogant and ridiculous! This stretch that would place a change in the mass of a moving object is also preposterous.

If one measures an energy level that conforms to an equation such as $e=mc^2$, or $e=\frac{1}{2}mv^2$, and decides that the mass instead of the velocity is the one that is changing, he must understand that the velocities are not the fixed parameters but can vary significantly. The changing of mass with velocity is a fabrication built on a false premise that seeks to assign limits where they don't exist in reality.

If one wishes to examine the equation $e=mc^2$ or $e=\frac{1}{2}mv^2$, one must understand that for a given energy level (actually measured or assumed), a photon would, in the first place, have to have a component of mass, and if that is so, that mass would have to go to negative infinity if the photon was at c! There is definitely something amiss with this logic.

Someone derives an equation, the equation is fundamentally limited in its scope and if moved beyond certain bounds, it is false; another person takes it and uses it to prove an absolute for universal motion!

The results are widely accepted by the physics community, and those who disagree and argue against this idea are held as defective in thought process; it is still false.

One of the resultant assumptions of much of this work then culminates in declaring that the laws of nature must follow the theory of relativity and the physical-time relationships defined using the Lorentz transforms, that they are not fixed but are covariant with the transformation process! How unbelievably arrogant!

We now tell nature how it must conform to our idea of what it should be doing with respect to a prime coordinate system defined by mankind! Once again, the height of arrogance! Finally we come down to a conclusion that would have us revisiting much that is thought to be known.

The variability of the speed of light, the immutable definition of time, and the lack of dimensional contraction as erroneously borne in the minds of some all lead to the conclusion that the known universe is not the size it has been estimated to be in the recent past, and its age is also not what some thought it to be.

Without the contraction, a new estimate would make it closer to half of its former self. The physical contraction (and subsequent expansion) due to a relative velocity is clearly not valid, nor is time dilation.

The Michael-Morley experiments found no "aether" in the cosmos, and the Lorentz transformations would be better termed the Lorentz fabrications.

Calculations and conclusions must adhere to the following *laws:*

1. The speed of light, c, is neither a physical nor an absolute velocity limit in the universe. Since c is not a constant, its use in equations as an absolute constant reference is therefore erroneous and is to be avoided, except for approximations, and should be noted thusly.

2. Every object in the cosmos has its own frame-of-reference system associated with the object at its origin and within which its movements and state of being can be defined.

3. Photons emitted from an emissive source move at c_{nom} with respect to the frame of reference of that source.

4. Photons emitted from an emissive source moving toward the observer will be detected by the observer as blue-shifted, as compared to that which an observer stationary in the reference frame of the source would see.

5. Photons emitted from an emissive source moving away from the observer will be detected by the observer as red-shifted, as compared to that which an observer stationary in the reference frame of the source would see.

6. Photons don't have a base color, frequency, or wavelength associated with them. They don't vibrate, nor do they oscillate to demonstrate a color. They portray no color information whatsoever but simply move at their velocity of emission v, where v_ε is the mean velocity of emission of the emitted photons.

The adventure continues.

WHAT IF?

What if:
 All photons are small particles (quanta) &
 All photons have mass &
 All photon mass is not the same &
 The speed of light is the constant?!

How about:
 All photons are small particles (quanta) &
 All photons have mass &
 All photon mass is the same &
 The speed of light is not constant?!

 Which one would you choose?

Let's look at the idea that the mass can be different. As a stream of photons would pass close by a massive star or planet, the path would deviate according to the traditional formula for gravitational force:

$$f = \frac{Gm1m2}{d^2}$$

where G is the gravitational force, the mass of the sun or planet is *m1*, that of the photons is *m2*, and d is the distance between the centers of mass.

If the mass of the individual photons varies, one would expect to observe a chromatic aberration associated with the stream that passes close enough to thus be affected.

The more photons of higher mass would tend to be drawn toward the massive body more than those of a lower mass. On the other hand, the more massive ones would have a higher momentum trying to keep them on course, while the less massive ones would have a higher propensity to stray since their momentum (forward) would be less.

If the photon mass is the same for all photons, then we would expect to see the slower photons move toward the massive body more readily than the faster ones. On the other hand, if there is a difference in velocity between photons of different energy levels (spectra), then one would expect to see a phase shift among the photons within the stream. In order to establish that, we would have to have a timed event such as a supernova, specific sun flares, or other photon pulsating source to study.

What would either of the aforementioned cases do in conjunction with the work Planck did with his black-body studies? We must realize, of course, that either case would not fit well with Einstein's work.

We must also derive and learn to write the spectral shift equations in an energy form instead of wavelength or velocity until the answer becomes clear.

It is clear and well understood that photon energy levels are inversely proportional to their wavelength, in that the energy level increases as the wavelength decreases.

It is similarly clearly and well understood that photons moving toward an observer from a source that is likewise moving toward the observer are seen to be blue-shifted, whereas those emitted by a source moving away will appear as red-shifted. This certainly seems to point to the photons of different velocities (relative) having color changes as well as energy changes. Perhaps the color content is simply a figment of the human optical system and not a color or wavelength at all! Do you suppose that the human eye with its multitude of photosensitive layers is actually sensing energy differences and converting that to various colors within our brains?

Perhaps color-blindness is actually energy-level blindness. Perhaps in cases of color-blindness, the multilayered retinal package has a short circuit from one layer to another so all colors register essentially the same, albeit with the possibility of varying shades of the base energy that the eye can perceive.

Let's go back to a sun emitting photons in every direction. As we sit there and watch the photons (and other radiation) stream from the solar surface out into space, we conclude that all of the photons moving in every direction all act like one another. They stream out at the velocity they are imbued with by the emission process within the solar mass. They have no idea that they have an oscillation, wavelength, or frequency component of any kind. They simply obey the force imparted to them and travel out on a path predetermined by their ejection process.

Now let us include a couple of colleagues in small spacecraft some light years away and stationary with respect to our position. They observe a photon spectral emission that is peculiar to this solar entity. Now let one of these craft approach at a high velocity; the observer now sees the same spectral pattern identifying our solar mass but it is shifted in its entirety toward the blue end of the spectrum. Now let the other craft move away at a similar high velocity and observe, again, the same spectral identity but now shifted toward the red end of the spectrum.

Well, what do we have then? We have a steady photon stream moving outward from our solar entity in every direction. We have an observer seeing a blue shift and another seeing a red shift within the same identical photon stream! It is not possible for the photons to have changed in mass, wavelength, frequency of oscillation, or any other way; they are simply moving on their vectored paths.

It becomes obvious that photon color is a function of their particulate mass and their velocity relative to the sensor/observer. If one used a sensor that was colorblind, he would simply measure a change in the energy level of the photon stream being observed. Again, we note that this is a function of the relative photon velocity with respect to the sensor and is independent of whether the source or observer is moving relative to the local space partition. The key is *relative* photon velocity. *But wait!*

Isn't the photon energy also a function of its mass? Of course it is. Now, if the streaming photons are all moving at precisely the same velocity relative to the emissive source then we have compounded the mystery and, instead of simply a velocity differential, we have also a momentum component as well. We recall that a particle moving with a higher momentum due either to its mass or its velocity has a higher energy level than the same one moving more slowly. Alas, we can't determine from this exercise so far whether the noted spectral change is simply due to velocity or momentum.

How can we resolve this issue? The only approach that comes immediately to mind is that of a pulsating broad-spectrum photon source that can be studied and quantified. We have observed such galactic sources in the form of novae, supernovae, and solar flares. Perhaps we should look anew at the possibility of a means to determine the true character of any spectral perturbation that may exist within such an occurrence.

If we watch a photon stream moving past a large massive body (huge planet or sun), and the emissive source is moving well beyond the massive body and coincidently moving behind it as well, what do we observe? The photon stream will curve around the huge body, as the emissive source in actuality has moved far enough so as to be completely obscured by the huge body.

The curvature of the photon beam has several items of interest; if the photons moving past are of various masses, i.e., a range of photon mass units within a given photon stream, we'll find the heavier ones more strongly attracted to that massive body.

They, however, have a higher momentum and want to remain on course, so is their path of curvature shallower that that of the photon of lesser mass? This also means they have a shorter distance to travel as they move past the planet. Perhaps the lighter ones will not be attracted toward the massive body as much, but their momentum is less, so they may have an identical path to that of the more massive photons.

Now as the photon stream moves past, it will reintegrate into the stream seen before its close brush with disaster. The lighter photons move back together with the more massive ones as if they had never parted. All back in their previous juxtapositions with one another. Nothing has changed.

An observer far to the exit side of this near collision will not see a phase shift of any sort and will simply notice a deviation in the photon stream path. As the emissive source now moves out from behind the large and massive body, the photon stream will appear to then jump forward, seemingly in an effort to catch up and then precede the emergence in a proper clarion manner. It will appear as if the large and proximal massive body is not large enough to completely obscure the emissive body in its transit behind, but this defies all calculations since the relative size as seen at the observer's position clearly indicates the opposite is true. It is simply a photon jump, nothing more, and it does not answer our momentum question at all.

Now, we understand that when supernovas and the like are seen, the astronomers have found that the released particles and radiations have reached us at different times.

Indeed the images produced in capturing the various radiations and photons show not only a difference in transit time but also a displacement seen only due to the temporal lag time. The class A-1 mentioned by our friend Dr. Zweig is such a case and is one from which we can glean more information than previously thought.

We may want to pause here for a while and have a closer look at what we call radiation and how the various forms are classified, their characteristics, and their domains.

We have gamma rays, X-rays (from soft to hard), light (photons?), neutrons, electrons, ions, beta, alpha, etc. Some are from atomic reactions, some are from high-energy events, some are casual in nature, and some are from manmade events and manufactured isotope emissions. Some of these are called different things because of the manner in which they are generated.

Much of the names and methods of cataloging of such occurrences are simply leftovers from the past and have not been upgraded to align with their characteristics instead of the source of their emissions, as should be the case. There are perhaps a goodly number of misnomers due to the problems with accurately identifying the source of the emissions; indeed, many would be so included.

Let's look at some photons and their origins, for instance. Some are the result of the heating up of a material that then releases the photons as a result of the thermal excitation. Some are given off as a result of nuclear events within the electron shells of atoms.

Some are the result of cataclysmic explosions. Some are from combustion of other matter. Some are given off by chemical reactions, and still others are a result of secondary particulate bombardment of other matter. All may certainly be classified as photons; why not X and gamma radiations and perhaps others as well? Radio waves come to mind, as do all manner of radiation within the envelope of electromagnetic radiation. A review of the methodology used to categorize the various known radiations is certainly a good idea.

Once again, we must be cautious to correctly categorize the items being considered as to their true and quantifiable characteristics. For many years, light has been included within the electromagnetic realm of radiations, and wrongly so. This misidentification was primarily due to the similar character of the waves of light and the inconvenience to those carrying on the investigations to delve further into the true nature of the light, i.e., photons acting in concert in a wavelike manner.

While much of the mathematical basis for electromagnetic radiation seems to fit the light waves as well, they are not the same, and closer looks at the metrics involved will soon show the apparent differences via specific experimental inconsistencies. We'll delve into this topic in another chapter since it moves substantially away from the point of our chapter on photons.

Let's take a closer look at the work of Planck and Einstein in the area of the photoelectric effect and energy relationships. Planck attached the energy of light to the wavelength λ, or $(1/f)$ of the observed light, where f is the frequency of the light oscillation. Einstein looked at the photoelectron emission of a metallic plate being bombarded by the photons and found that the electrons were emitted in discreet energy steps that were proportional to the light wavelengths and not the light intensity. He concluded that light was made up of quanta and that the energy of this light (certainly now to be considered quanta) was grouped in "packets" of discreet energy levels and, therefore, was indicative of quanta and not waves!

As he had done on several occasions, Einstein misinterpreted the results and carried the discreet energy tag over to the photons and therefore declared them to be quanta. In fact, the photoelectrons demonstrating the clear step function in their energy distribution is simply a result of the step function of the electrons being released from the host atoms in discreet energy steps. The energy of the photons will still change as a relationship to their frequency or wavelength in a smoothly continuous manner throughout their entire range of existence. The steps of photon energy distribution are a continuous function, as is their specific mass or state of relative velocity with respect to a given observer.

The conclusion was correct, but the causal factor was misinterpreted by Einstein. This fundamental propensity of Einstein's to lay the rationale of the resultant observed action onto the causal effect is disturbing and leads to erroneous conclusions if analyzing the results of experimental data.

Consider the pebble being dropped from the train as it moved along the track in 1905. He saw the pebble moving in different paths depending on the perspective from which it was viewed. The pebble, however, had only one path of motion. It never changed no matter from whence it was observed.

Again from 1905, we see a mass-energy equivalence being proposed. Perhaps it's time to firmly and clearly define terms we use so as to avoid confusion. What is mass? What is energy? What is momentum? What is velocity (speed)? Can we define these terms while not attaching them to one another?

Let's look at mass first. Mass is the fundamental condition of existence or being. If a particle, no matter how small, exists, it has mass. If an object comprises sufficient numbers of a variety of particles in a seemingly coherent group such as a baseball, that object has mass (which is simply a summation of all of the various particles of which it is formed).

The ability or inability of modern science to measure the mass of a particular particle neither proves nor disproves its existence; thus, mass may very well not be a measurable quantity but must simply be inferred by calculations using data gathered experimentally.

Let's look at energy. Energy is both a function of a mass in relative motion as well as the cause of a mass to be in motion. It must be defined as a salient property of the mass of an object. This does not make it equivalent to the object itself but simply a property of it. Energy insinuates movement of the mass or the requirement or the application of a force acting on the mass, thus transferring an amount of *energy* to the mass causing it to move.

One might say that the energy attributed to the photon mass (kinetic energy) is equivalent to the energy of this mass at rest. Further external energy supplied through application of an applied force simply accelerates the mass and thereby becomes an attribute of it as the photon is slowed or sped up in its relative velocity or has its path of flight changed.

Some speculation as to the at-rest mass and kinetic mass of a photon is different. One must again ask with respect to what frame of reference is this judged? Is the photon at rest in any frame of reference other than its own? We have seen that photons are moving about the cosmos at varying velocities that can certainly be from zero to high relativistic velocities, exceeding the popular c proposed as the astronomical speed limit. How, then, are we to judge the rest versus kinetic mass of the quantum? So far we have a property of a condition of being! Somewhat esoteric? Nothing here indicates, nor can it, that energy is the same as mass! There is no equivalence. One is the mass, and the other is a state of that mass.

As much as some people would like to use systems wherein $c = \lambda v$ (v is also typically known as f for frequency), we understand that the length of a wave of a given frequency is given by $\lambda = 1/f$, which brings $c = \lambda v$ to $c = 1$! An absurd identity with meaningless function mathematically and in the physical world.

Let's look at momentum for a moment. Classically, we have $\varrho = mv$. This equation is not correct without inclusion of the idea that v is strictly a relative term and has no absolute value. The idea of the momentum of a particle or object of mass requires the observer to identify the value of its velocity with respect to his own particular frame of reference.

The momentum of a mass is, therefore, unique to that observer and his frame of reference, and, if related to another colleague, the information must include a transform that identifies the conditions of relativity between their two reference systems as well.

We decide to tell a colleague who is cruising about in a spacecraft while we ourselves are in another; we're both off into the vastness of the cosmos, and a small meteor passes by us at high velocity. We can't tell our colleague that it was traveling at X miles per hour because that has no meaning to him unless we both can agree that his reference system of coordinates is fixed to ours or we can define the movement of his reference system with respect to ours.

On Earth, if we say that an aircraft is moving at 537 miles per hour, we are automatically inferring a reference to the surface of our planet. That has no relevance to some extraterrestrial colleague that hasn't even seen the earth (or has any idea what a mile or an hour is). Again, we see these ideas as strictly relative terms, and what we see is not necessarily what a colleague sees or would measure from his perspective. Remember the pebble dropped from the rail car. Who is watching and from where?

This gives us a chance to look at another difficulty that has bothered many scientists for a century of so. If we are resident near a solar orb and it is emitting photons in every direction equally well, even though those photons have either of a variety of different velocities or mass, they all appear the same to us, observing from nearby.

Now we have two observers some distance away, and they are moving with respect to us; one approaching us, and the other is receding. Perhaps it is us who are moving?

Again, we need to have a reference system as the prime system for discussions. We'll use ours. Here's the question: do the observers see our photons as more or less massive? If we were emitting rods, would the observers see them as different lengths? For the observer moving smartly toward us, do the rods shorten in length? Of course, the answer to all is *no*! There is no such thing a physical contraction due to high velocity of motion.

Hendrik Lorentz made some wild assumptions and mathematical gyrations trying to prove a null result of Michaelson and Morley's experimental data. The Lorentz transformations produce a fiction that is tempting for the non-thinking student.

Yet another result of this experiment is that the two observers see our photons as having shifted, toward the red end of the spectrum for the observer from whom we are moving away, and toward the blue for the observer toward whom we are traveling. Notice that our photons are simply traveling at the constant velocity at which they were emitted from our solar orb (unless they get acceleration due to action by an outside force). They are simply being perceived as having changed in spectral content. This is simply a change in the relative velocity with respect to the observer's frame of reference and not a change in the photons themselves.

Just for amusement, there is no such thing a time dilation, either: no space-time continuum, no stretching of space fabric, no threads, no strings, no wormholes, and no parallel universes. There are no matter transporters and no time travel.

Objects in the cosmos can and do travel faster than the speed of light (c) and, by the way, with respect to which frame of reference do we measure these photons, anyway? The velocity of our photons perceived by the distant observers does most certainly change, as witnessed by the perceived spectral shifts. The velocity of the photons is fixed with respect to their source of emission. The observers see $c +/- \Delta v$, where Δv represents the approach or receding velocities between the source and the observer's position.

Ergo, the velocity of light is referenced to the emitter frame of reference and is not a constant, neither universal nor fixed.

$E \neq mc^2$! At best, the equation must be, at the very least, more precisely ordered to $e = m(c + v_{rel})^2 k$, where Δv_{rel} is the photon velocity differential with respect to the observer's location, and k is a proportional constant.

Let's look at another way to make this differentiation; five solar orbs are all seen together and four are traveling in the +X direction with respect to the local space partition (LSP). All of the orbs are emitting their photons, and they are also moving at different velocities.

For a casual observer some distance away, the photons from the different suns are seen as significantly different. The four that are traveling are shifted toward the blue end of the spectrum, some more than others.

Which are the true photons? Which solar frame of reference is the preferred frame of reference? Which photons are moving at c? Doesn't work, does it? Again, for those who didn't hear it the first time:

$E \neq mc^2$!

Clearly, we have now shown that photons can and do indeed travel at a range of velocities. Is there any way to determine whether the mass of all photons is identical to all the others? It would seem that since the photon is a minute fundamental subatomic particle, it is unlikely that we'll find variations in their mass, one to the next. The variations in velocity can account for the spectral shifts that have been found; this by itself may prove sufficient to discount the probability of any variation in particle mass.

One approach perhaps would be the determination of variations in photon energy levels with changes in relative velocity. The accumulated data on photon energy, along with the various ideas put forth by researchers, is voluminous and perhaps is adequate to provide a benchmark from which we can expand our hypothesis.

Most of the work has steadfastly continued Einstein's idea that all photons move throughout the universe at a constant speed, c. With this invariable, they then related the photon's energy content to its wavelength, λ, (or frequency, which is given as $1/\lambda$).

Well, we have proven the idea of range of velocities for photons and the consequential result of a variation in both apparent spectral and energy content. In order for the scene to become completely clear, we can move to strike the idea of the photon frequency of oscillation, or wavelength characteristics. The simple fact is that humans have assigned a color representation to the photons as a carryover from classical optical studies from centuries ago.

We have taken our human ocular receptors in their limited form and used their capabilities as the benchmark for all perceptions and, thus, the character of observed phenomena. The eyes of biological creatures are varied in their manifestations and configurations.

There are those that are highly multiplexed in bringing their images together and some that are thought to possess limited chromatic response. Some are highly sensitive to ambient light, and some welcome augmented conditions using artificial sources. Perhaps it's time once again to ask, "What if?"

What if: One more time, the biologic ocular receptor (eye) doesn't really see color, per se, but simply uses an ability to carefully discriminate between variations in photon energy levels. The brain, in concert with a multilayered retinal target, captured the impinging photons and, by determining their strength of impact upon the photoreceptors in the retina, determined what color is being seen.

Perhaps the condition of color-blindness is a condition of the inability of the retinal package to effectively discriminate this differential. A study of the signals sent through the optic nerve to the brain would aid in understanding this concept as valid or not. This makes one think that color reception in colorblind individuals might be treated by enhancing the sensitivity of the retinal photoreceptors to minute differences in energy levels.

If it is found that color is simply a manifestation of the human vision system and does not, in fact, describe the oncoming light characteristics as having a frequency of oscillation or wavelength, we can begin the study of photons with a new approach.

Of course, this colors past work in erroneously categorizing light within the electromagnetic wavelength spectrum. Perhaps we can consider this: light may be treated as waves in considering how it acts and reacts, but photons are indeed particles that have mass and an energy component based on that mass and their velocities with respect to the observer conducting the experiment.

Considering photons as the very basis of electromagnetism is perhaps one of the great fallacies that has been carried on throughout the past decades of research. The work done by Maxwell, et al, in electromagnetic wave characteristics and theory was carried over to the study of light because of its wavelike behavior.

The simple fact is that the photon stream with its extreme number of constituent quanta behaves in a wavelike manner, in much the same way as a stream of water molecules. While not exactly analogous, many of the behavioral characteristics lend themselves well to both applications. One brings to mind experiments such as the dual slit experiment producing interference fringes, etc.

Now, we take on a side question that asks: if all photons have the same mass and their kinetic energy levels vary simply in proportion to their respective velocities, what changes in velocity correspond to the range of energies already known to have been exhibited by photons in motion?

We have clearly defined range of energies already observed (and this is not to say that this comprises the entire range by any means). We could use the Doppler equations that have been derived, molded, and massaged to conform to observed events, or we could use Planck, as well. Using Planck will certainly give a close approximation to the true answer; perhaps we could try both and then compare the results (or discrepancies)?

Planck said that e=hf, where h is Plank's constant and f is the frequency. We could use the λ, which is simply $1/f$, but what about velocity? What about the mass component? We could use other identities for e, such as $\frac{1}{2}mv^2$, or mc^2, or $m\Delta\upsilon_p^2$. The problem here is that we don't know the mass, and we don't know the velocity.

No, we won't rely on the older derivations of these but must rely on some of the practical results of actual experimentation to provide more concrete numbers.

If h, f, λ, and e are all easily found, how do we go about finding m or v?

Let's restrict our present efforts to that region visible to the human eyes, typically placed between 380 and 740 nanometers of wavelength. If we look at the mechanism termed the photoelectric effect, wherein electrons are released from their atomic bond by the action of impinging photons of a particular energy level (wavelength), we find a number of scientists who assume that a particular number of photons are needed to effect this electron release. They then determine that the electron mass divided by that number of photons will yield the value of the mass of a photon.

One must ask exactly how many photons of the needed wavelength are required to release one photoelectron. Perhaps just one? Ten? Eighteen? How do you count just one photon?

No doubt, there are more questions than there are answers. If we could even accurately count each photoelectron being released, we still would have to make assumptions in order to determine the number of photons involved in accomplishing this task. Far from conclusive, to say the least.

RED SHIFT? REALLY??

RED SHIFT, REALLY?!

They're at it again. Some years ago (about fifteen), a dedicated scientist did some calculations to try and describe as accurately as he could the source of some gamma radiation bursts that have been detected. These bursts were of such a magnitude that most cosmologists thought they were from a super massive source on the other side of the galaxy from us (but still within the bounds of the Milky Way).

The calculations and some subsequent data began to indicate that these bursts were either so unbelievably massive or the poor scientist had made errors in his conclusions, and $e=mc^2$ was still safe from being exposed as erroneous (at last).

Further probes were sent to detect these bursts, and the results were troubling beyond belief; the scientist had been right all along, and the cosmology community was in a desperate uproar trying to save Einstein (and themselves, by the way). These gamma bursts were far beyond explanation with any current thinking, and they needed to come up with some theory that would slip the results into a comfortable niche and not rock the boat further.

Finally they came up with it (!?). All of a sudden, the idea that the red-shift associated with the bursts proved that the sources of these bursts were located at a closer distance than previously thought, and therefore the source size need not be so massive after all.

MATH-A-MAGICIAN

They (and Einstein) were saved! But *wait*! It seems to me that we have gone over this ground thoroughly, and even using the previously accepted theories we understood (or thought we did), determined that the red spectral shift indicated a relative velocity of the photon source to us (the observers).

A receding source meant that the source would have a spectral shift toward the red end of the spectrum. Yes, we do remember that. Yes, this was the prevailing theory.

Now all of a sudden, it means that the source is at a particular distance from us and not that it is moving away! The Δv (changing velocity) has now turned into a distance! First we have mass turned into energy and energy to mass, and now we have velocity turned into distance; how interesting.

But you must understand that this entire exercise is to align with the status quo and save Einstein and e=mc². Case closed, and all good fellows shall get on this bandwagon and fully subscribe to this new approach.

Really, fellows, do you really expect us to buy this material? Perhaps it should more appropriately be used for fertilizer. It's time to quit trying to make the data fit the theories and simply formulate the theories to fit the data.

Perhaps a reminder that upon viewing the "deep field" by the Hubble Telescope, the galaxies visualized thereby were found to be imbued with a variety of spectra in a wonderfully random array and strongly in contention with the ridiculous "red color equals distance" theory brought to the table by our previously distinguished colleagues.

The bursts are coming from all over the cosmos, and they are so massive that no known source has either the proximity or mass sufficient to have produced such results. It's time to back up and try again.

FORCING A THEORY?

If huge (meaning unbelievably massive) celestial bodies exist at some far off distance and they are evidently the source for such bursts, we should be working on trying to understand this as the probability, trying to define such bodies, and trying to expand our vision about the universe we live in, not trying to bend the rules to cover our collective egos.

Further energy considerations are required and must include atomic interactions and reactions apart from the pedestrian energy associated with particulate momentum. When looking at the various explosive occasions throughout the universe and, in particular, novae and super-massive subatomic cataclysmic eruptions found throughout the universal continuum, we must then move toward including the strong atomic force and subatomic particulate interactions. These are apart from and wholly additional to the $e=mc^2$ interactions discussed thus far.

In examining the data gathered by our gamma detection satellites and watching the cosmos surrounding us, we find the massive explosive events far surpass the incorrect and less significant $e=mc^2$ characterization of particulate physics.

The ubiquitous presence of atoms and molecules in such a vast volume can contribute to explosive occurrences far beyond the simplified "mass at some velocity" consideration, thus far allowed by our esteemed astrophysics community.

This, of course, leads to the acceptance that gamma bursts of such extreme quantities of radiation must include this subatomic force interaction and therefore be a result of that interaction.

Shed your traditions and awaken to a new approach that seeks to truly understand the workings of the universe.

Keep looking, analyzing, formulating, and learning, but don't ever believe you have *the* answer. You probably only have a part of the story. The rest of the story is yet to come, perhaps tomorrow, maybe next year, maybe…

BIBLIOGRAPHY

Relativity Unraveled: A Question of Time, Hans J. Zweig (2004)

Philosophiæ Naturalis Principia Mathematica, Isaac Newton (1687)

On the Law of Distribution of Energy in the Normal Spectrum, Max Planck (1901)

Electrodynamics Phenomena in a System Moving With Any Velocity Smaller Than That of Light, Hendrick Lorentz (1904)

Special Theory of Relativity, Albert Einstein (1905)

Special Theory of Relativity, Albert Einstein (1917)

United States National Aeronautics and Space Administration, et al. (numerous reports and papers on astrological data gathering whose voluminous quantity precludes specific acknowledgement)

APPENDIX

APPARATUS FOR MEASURING THE SPEED OF LIGHT (IN A LABORATORY ENVIRONMENT)

APPARATUS FOR MEASURING VARIATIONS IN "*c*" AS A RESULT OF CHANGES IN WAVELENGTH AND ATMOSPHERIC COMPOSITION

W.J. McKee

© 4/12/2006

ABSTRACT

In following the various experiments, both the "rigorous" and "thought" varieties, and perusing the numerous volumes of written material on the subject of the speed of light, in view of the large amount of work that has been done and will follow hence, I have designed an apparatus that will aid in the identification of whatever variations may be found in the speed of light with respect to differences in wavelength and environmental pressure, as well as the types of gas molecules that the light stream might encounter on its journey about the universe. The result is the apparatus description and schematic that follows. Finally, a definitive experiment and suitable apparatus to accomplish it can be undertaken. The resolution of the apparatus is in the 1×10^{-8} **c** range; not as good as the Lunar

Laser Ranging system at University of Texas, but with the additional ability to vary the wavelength along with light path atmospheric characteristics, it should provide good results.

DESCRIPTION OF THE APPARATUS

The apparatus is designed using some of the most modern technologies available to us, items that were not available to the great minds of the past. Many have built similar devices and, had the newer technology been available to them, better results could have been obtained.

A high lumen output white light source is utilized. An adjustable monochrometer is included immediately following the light source. Slit apertures are used, as is an objective optic, the output of which illuminates the source slit aperture. A rotating mirror assembly is used. It is fitted with a two front surface mirrors used for light reflection of both the timing and main light beams. A stationary mirror assembly is also provided to allow crude main beam adjustment. This mirror assembly, along with the smaller beam path length extender mirror, also acts in lengthening the beam path length. Finally, a sensor array is used to detect the impinging light beam at its final destination. The apparatus is built largely upon a rigid optical bench, and the entire apparatus, with bench, is housed in a high-integrity environmental chamber.

THEORY OF OPERATION FOR THE APPARATUS

A high-intensity broadband light source (rich in the photopic wavelengths) is utilized since the *initial* efforts are constrained to that photopic region of the electromagnetic spectrum. A monochronometer is used to select the various wavelengths as desired. An objective optic is used for light beam collimation. A number of slit aperture plates are utilized at several key points along the light path to act as baffles and minimize stray light interference with the test results. A primary timing light source-and-sensor assembly is used to watch the rotating mirror assembly and provide a timing pulse for use in triggering the various electronic functions used. A second timing light source-and-sensor assembly is located on the opposite side of the rotating mirror assembly and is used to ascertain the precise speed of rotation of the mirror assembly.

As the light is brought to the main stationary mirror assembly by the rotating mirror assembly, it is "scanning" through a brief arc. This arc is subtended by a small concave front surface mirror as part of the main stationary mirror assembly and acts to maintain a light beam path direction into the light beam extender mirror group.

It is clear that the light pulse width will be on the order of 10 μs or so as the rotating mirror brings the light beam to the main stationary mirror assembly and sweeps across the small concave mirror. The pulse width is of little concern with respect to the experiments presently planned.

A long path length (D_1) is provided between the source slit aperture and the final contact with the rotating mirror assembly in order to provide a time delay between the initial timing mark and the subsequent detection of the signal at the final sensor array. The main stationary mirror assembly, along with a smaller beam path length extender mirror, is utilized to increase the effective path length of the light beam by way of numerous ping-pong style reflections; it also provides a coarse zero adjustment of the final beam traveling to the sensor array.

The sensor array utilized herein is based not on interference patterns or visual observation with a microscope but a linear photo sensor array containing a multiplicity of individual sensitive segments (1,024 in this case but can be otherwise).

The zero adjustment simply selects which sensor segment is the zero point for the beam landing. While this is somewhat arbitrary, a point within the central zone of the sensor is recommended to allow for any movement in the landing spot of the perturbed beam as wavelength, chamber pressures, and various gasses are introduced into the light beam path.

The rotating mirror assembly has multiple uses; it reflects the timing beams from the timing light sources to the timing light detectors. This then provides a reference pulse to trigger the electronics module that is used to manage the sensor array output information and secondarily provides a second pulse that will be compared, in time, with the primary timing pulse in order to precisely determine the rotational speed of the rotating mirror assembly. The next function is to reflect the main light beam to the main stationary mirror and also the secondary reflector in the light beam path length extender.

The final function for the rotating mirror assembly is to reflect the light beam coming from the beam extender onto the sensor array. The sweeping across of the beam onto the sensor surface provides a result that manifests itself as a sloped ramp from some null level to a high-level proportional to the number of photons captured. It is easily seen that the slope will have a sharply defined rise time if the source of the light beam is very monochromatic; a broader slope would be indicative of a photon beam that has a range of velocities. Movement of the signal to the left or right would indicate a decrease or increase in beam transit time.

Finally, the sensor array is cryogenically cooled to minimize the ambient noise and allow a high-gain amplifier to follow while maintaining a usable signal-to-noise (S/N) ratio.

A computer is used with appropriate attachments to manage and watch all functions of the system under test. Algorithms are included within the software to gather the sensor data, calculate the speed differentials, and monitor the specific wavelength, chamber pressure, molecular content, and temperature within the chamber.

Utilizing the computer, it is possible to determine the difference in time between the initial timing pulse and the secondary one. With this information, the precise speed of revolution of the motor is determined and applied to the algorithm used to calculate the actual speed of transition of the light beam through the apparatus and then onto the sensor array.

Some speculation as to the relative speed of light as a function of wavelength has been suggested with the proposition that the various wavelengths of the electromagnetic spectrum (photons) travel at different rates of speed. If this is correct, we will be able to determine the differences with this apparatus. If a monochromatic light source is used, it is expected that the sensor will show a rise time of a segment or two for the incoming pulse of light. If, however, the source is rich in a variety of wavelengths, the pulse will show itself as gradually rising to peak value through a number of segments. A good test may be to introduce a broad-spectrum white source and then compare it to beams from the far end of the photopic spectrum.

It is certainly viable to utilize the apparatus for wavelengths through the ultraviolet and infrared portions of the spectrum as well. A similar apparatus with modifications can be adapted to both the X-ray and radio portions of the spectrum as well. Attention must be paid to the sensors selected for such applications, as well as the lower-level sensitivity and ambient noise level from the sensor as presented to the following buffer-amplifier.

SUPPORTING CALCULATIONS

According to most sources, the speed of light is purported to be 300,000,000 meters per second (3×10^8 M/s). Converting to microseconds for convenience in the following visualization, we have 3 x 10^2 M/µs.

We'll suppose that the motor and rotating mirror assembly is rotating at a speed of 3,600 revolutions per minute (RPM) or 60 revolutions per second (RPS). For a light beam path length from the rotating mirror assembly to the sensor array of, say, three meters (D_2), and a sensor length of one centimeter (cm). It's clear, therefore, that there are 100 sensor lengths per meter length on the sweeping arc of the light beam being measured.

We want to determine the sweep time across a single sensor, and we find that the sweep time of the light beam across the sensor (Tss) is given by:

(1) $Tss = (2\pi \cdot D_2 \cdot RPS \cdot N)^{-1}$

Or $= (6.28 \cdot 3 \cdot 60 \cdot 100\ )^{-1} \approx 8.84\ \mu s.$

where D_2 is the beam path length between the rotating mirror assembly and the sensor array, RPS is 60 revolutions per second, and N is the number of sensors that would fit within a meter length (in this exercise, there are 100 sensors per meter).

Now, for each sensor segment (1,024 segments per sensor) over the sensor scanned length, we then have segment scan time (T_{seg}), which can also be regarded as the limiting resolution of the sensor given by:

(2) $T_{seg} = (Tss\ /\ N_{seg})$

or $= (8.84\ x\ 10^{-6}\ /\ 1024) \approx 9\ x\ 10^{-9}\ sec$

At a base light speed (c) of 3 x 10^8 M/s, which is equivalent to 3.33 x 10^{-9} seconds per meter (1/c), and with a segment scanning time of 9 x 10⁻⁹ seconds, we find the system resolution (R_{sys}) from the following:

(3) $R_{sys} = T_{seg} \cdot c$

Or $= (9 \times 10^{-9} \text{ sec} \cdot 3 \times 10^8 \text{ M/s}) \approx 2.7 \text{ M}$

This means that we have the ability to detect (or have a system resolution of) a speed-of-light difference of +/- 2.7 meters per second. Certainly the motor scanner speed can be changed to provide a resolution that the operator may find more desirable or convenient for whatever reason. One limiting factor here is the number of photons that are emitted by the source, and that can be captured by the sensor array at the scanning speed desired; this, along with the electronics S/N ratio and electronic system bandwidth require careful selection and design.

The apparatus resolution is not as good as the lunar laser ranging system, but this device will afford us the opportunity to vary the wavelength and the atmospheric conditions through which the light beam must travel.

Not shown but essential to the construction are the optical bench and the environmental chamber within which the apparatus is housed. The chamber is capable of both high positive pressures (up to approximately 100 PSI), as well as high vacuum (down to approximately 1 x 10⁻⁹ Torr or better). It is essential that a clean vacuum system be used; no diffusion pumps or mechanical roughing pumps should be used due to the contamination they bring to the vacuum chamber. Cryogenic pumps and ion pumps are preferred for this effort. While this is not a perfect vacuum, it will certainly suffice to bring some first-order effect information to the community. It is clear that the ability to draw down a vacuum and then backfill with a gas or a gaseous mixture up to a wide range of pressures is possible. The chamber will provide a capability for temperature and humidity control along with the pressure variation.

The ability to vary the molecular component, the chamber pressure, and the temperature, along with the ability to choose whatever wavelength of light we wish to measure, provides an optimal test, is exceptionally valuable, and has not been done heretofore.

CONCLUSION

It is clear that the use of modern technology, along with the apparatus design, as shown attached, will finally result in a firm definition of the speed(s) of light. After many years of supposition, thought experiments, data gathering, speculation, and guesswork, we can come to an agreement supported by empirical data from a series of carefully designed and well-monitored experiments. It is anticipated that future apparatus additions may also include refinements such as a high-gauss magnetic field sector in the light beam path, as well as an electrostatic field transition sector, to further our knowledge of the characteristics of the beam reaction to such external forces. It is also very desirable to incorporate ion traps, as well, to allow measurements of any ion generation/excitation that may be found as a result of the introduction of some of the gas atoms or gaseous molecules in the beam path.

The reader is referred to the accompanying drawing, entitled "Speed of Light Test Apparatus" for an annotated schematic of the basic apparatus.

RELATIVE "SPEED" (RELATIVE TO "C")	"WAVELENGTH" RECEDING 552 NANOMETERS -STATIC	"WAVELENGTH" APPROACHING 553 NANOMETERS -STATIC
"C" +/- ↓		
0.005	552.75	547.25
0.01	555.5	544.5
0.02	561	539
0.04	572	528
0.06	583	517
0.08	594	506
0.1	605	495
0.2	660	440
0.3	715	385
0.4	770	330
0.5	825	275
0.6	880	220
0.7	935	165
0.8	990	110
0.9	1045	55
1	1100	0

SPEED OF LIGHT MEASURING APPARATUS

REFERENCES:

1. A. Einstein, *Relativity, The Special and General Theory*, 1920
2. H. Zweig, *Relativity Unraveled, A Question of Time*, 2004
3. S. Schleif, *What is the Experimental Basis of the Special Relativity Theory*, 1998
4. I. Stavinsky, *Flaws in the Logic of Einstein's Special Theory of Relativity,* 2000
5. R. Polishchuk, *Derivation of the Lorentz Transformations, 2007*

The author humbly wishes to acknowledge the large volume of work done in the areas of relativistic, particulate, and astrophysics by the great minds that have gone before, those that remain and persevere, and all those that are unafraid to question.

INDEX

INDEX

151

ABOUT THE AUTHOR

AUTHOR

Mr. McKee is retired from a career of a broad technical character. His formal training was in Electronics Technology, which was followed by teaching, design engineering, radiation imaging, and finally 3D imaging system design. Most of his patents are in optical systems, engineering, and applications. He, like many of his colleagues, has an interest in and questions about relativity, photon physics, and the cosmos